How to Start a Vineyard in 2020

By: Alex Johnson

Table of Contents

Introduction

This guide is your tool for learning the vineyard business and creating a profitable winery from the start. The guide will teach you everything you need to know about grapes, how to make wine, how to sell wine, and business strategies specific to the industry.

The first portion of the book will teach you the language of wine. Reading through the terms will give you a good foundation before getting into other topics of the book. Learning some of the most important terms will give you a better understanding of wine and increase your professionalism. Throughout the process you may find yourself talking to other interested parties about your business, or other professionals who will expect you to know correct terminology. Although the list is only scratching the surface, it's a good start and the most important terms are taught.

Competition, timelines, and costs will then be briefly discussed. The purpose is to educate yourself and know what you're signing up for. A winery can be a costly investment and knowing the basics from the start will prepare you for what's ahead. The vineyard business has tight competition and may be even tighter in your area. Understanding this will help you form your business to be more unique and stand out. Starting a vineyard takes time and money. Patience and understanding the timeline of profitability is important. The book will touch on ballpark costs but will not be very specific to your region. Having an idea of what to set aside cost wise, is a good idea.

The first part of the book will also touch on creating a business plan although there are other relevant chapters pertaining to this topic. The goal of the first chapter will be to create a good foundation of knowledge before making decisions. A checklist will be given to help you get started.

The second chapter will discuss selecting a site. Depending on your region you may have limited options but in the wine world, winemakers know that each site is different. There are many different components to a good site. This guide will teach you all of the components that will affect the final quality of the wine. What many don't realize is that the decisions made throughout the process all directly affect the quality of the wine. In Chapter Two, you will see how important each decision is going forward. Land elements, climate, soil, global positioning, and vineyard positioning will be explained. After reading this chapter, you'll feel more comfortable selecting a site and knowing how each component will affect your final product. Chapter Two will also give you the basic knowledge needed to design your vineyard and trellis system once you have a site.

The book will then move into information about the grapes themself. Considering you're likely hosing a site and starting your business in the United States, brief introductions to the regions will be given. As you know, the United States has a variety of weather patterns and subclimates. Many differences in terroir (or the natural environment where a particular grape/wine is produced) will be impacted by regional location. This section will be helpful for giving you a place to start. Looking at the success in your region will give you better ideas for your winery. Don't feel like you're copying a region because the decisions the winemaker has impact the wine the most. However, studying varieties, soil techniques, and copious weather decisions will set you on the right path.

The same concept can be used for growing different varieties in specific regions. Information on the most populous grapes will be given to help you decide what to grow. Chances are you have an idea of your favored variety. This portion of the book will discuss the basic information on the most populous varieties and their favored climates. You will learn that many climates are able to grow grapes, but the climate will play a large factor on the grape's success yieldwise and quality wise.

Chapter Four will then take you through the winemaking process from start to finish. All of the potential steps will be touched on to give you the basic knowledge necessary for getting started. Many don't know how wine is made and all of the steps required. This section will give you useful information, and lay out hard to find information. The chapter will dive a bit deeper on many of the steps and help you make better decisions for your business and wallet. It's possible you may enlist professional help during these processes when getting started, but the basics are crucial for running a successful winery.

After learning everything you need to know to get started, the book will move into practices for after the wine is made. So often it's more about selling the wine than actually producing it. Of course quality is important but even a quality bottle can have trouble selling if the right business and marketing practices aren't put in place. You will learn how to market your final product and receive suggestions on different techniques. In addition, a full section will be dedicated to digital marketing. This includes information on how to get your website noticed and setting up social media sites.

The last part of the book will discuss accounting and tax opportunities. This section of the book will introduce you to financially managing your business and show you where to put your energy. Tax opportunities will be discussed but it's important to remember that your situation will be unique to the location where you're residing.

After reading this book you will have everything you need to start and run a successful winery business. Not only will you learn about the wine industry itself, but you'll receive insider tips and strategies to help your business take off! We'll teach you everything you need to know and the do's and don'ts in regards to this business. You'll learn the most common mistakes wineries made - and, most importantly, how to avoid them! If you're ready to learn how to create a profitable wine business, keep reading!

Chapter 1: The Basics

Getting into the wine business can be a great decision. Whether starting from scratch or taking over established land, the possibilities in the wine industry are endless. The demand for wine is unwavering and is one of the most popular beverages of all time. Wine is everywhere and has been around for ages.

Winemaking is a multifaceted business with many different components. In today's world, vineyard tours and wine tastings have become experiences that the public can't get enough of. In the world of social media, one enjoyable experience can lead to many. The interest surrounding wine is ever growing and won't slow down any time soon. Wine connoisseurs are planning lavish getaways to visit local and destination wineries, with some of these visitors making huge wine purchases.

For a winemaker, providing excellent wines and great experiences is the double-appeal for a successful business. With all that being said, there are some things to consider before jumping with two feet first into a complex wine business and industry. Understanding the competition, the timeline of a successful vineyard, and all the necessary costs will determine the outcome of your business. The information discussed in this chapter isn't meant to deter you, but rather to give the whole picture so that you know what to expect with little to no surprises. Wineries require a lot of work, time, and effort but also have the potential to turn a major profit.

Wine Lingo and Glossary

If you've ever been to a wine tasting or sat down with an experienced vintner, there may have been a few unfamiliar words being tossed around. If you were left scratching your head, don't worry, you're not alone. Wine terminology could be described as a whole new language. Although this book will use simple terminology, it's important to know the correct lingo that should be used as you gain experience. Although there are over a hundred terms to be learned, this section will go over the most important. The terms will be given in alphabetic order. Understanding these terms will help you produce better wine and help you be better informed if things go wrong.

1. Acidity

Acidity can be described as fresh, tart, herbaceous, or sour tastes that originate from the natural fruit acids found in wine. Tartaric, malic, and citric acids are the most commonly found in wine. If a wine is described as "crisp" it's often because of the high acidic levels. If the wine contains too much acidity it may be referred to as "green". If a wine is described as "flabby" this means the wine lacks necessary acidity. Wine can evoke a mouth-watering sensation, similar to drinking lemonade, which is caused by its acidic nature. Acidity in wine is necessary to balance sweet and bitter sensations. The acid in wine also helps to age and preserve wine longer. Each variety of wine will have different amounts to properly round out the final taste. Most wines will range on the acidic side of the pH spectrum with most ranging from 2.5 to 4.5 (Puckette, 2019). Total acidity listed on bottles will note the overall acid concentrate while the pH value explains the intensity in regards to taste. Unripe grapes tend to have a higher acidity than ripe grapes. Sweetness occurs as a grape ripens. The best moment to harvest is when the balance between sugar and acid is compatible for making wine. Shorter growing seasons or cool, nighttime temperatures typically produced wines with high acidity. Cool weather or grapes that don't fully ripen are usually the most acidic.

2. Aroma vs. Bouquet

The term bouquet takes into account all fragrances, smells, odors, or scents a wine illuminates. Bouquet is different then the term aroma which is also used when describing wine. Wine aroma refers to a grape variety while wine bouquet refers to the winemaking process, fermentation, and aging. Examples of terms associated with aroma are fruit, flower, or herbal scents. Examples of terms associated with wine bouquet are brown sugar, vanilla, hazelnut, cigars, spices, and tobacco notes which are derived from the fermentation and aging process. Aging wine will alter the original aroma compounds after fermentation.

3. AVA

AVA or American Viticultural Area refers to a region in America that has been approved by the Bureau of Alcohol, Tobacco, and Firearms. States can range anywhere from two AVA's to over one hundred in areas like California.

4. Balance

The term balance takes into account alcoholic strength, acidity, residual sugar, water, and tannins. When all are in perfect harmony, a wine is said to be balanced. If one element overpowers the other, an inappropriate ratio occurs and is picked up by an experienced wine drinker. Balance plays a huge role in the final taste, finish, and quality a wine has.

5. Blending

Some winemakers chose to blend different varieties of wine. Blending is helpful to add more flavor to wine, mask inconsistencies, or to create a new style unique to the specific winery.

6. Body

This term may be more familiar as many bottles have full-bodied or light-bodied written across the label. Body is the term for how heavy the wine feels on the tongue and can also take into account texture. Alcohol, extract, glycerol, and acid are the elements that determine the body of a wine. The region in which the wine is produced will also play a factor depending on how high the sugar-alcohol content gets when the grapes ripen. The warmer the climate, the higher the potential for extremely ripe, high sugar grapes. Because alcohol content contributes to the viscosity of wine, it can be assumed that the higher the alcohol content, the more medium to full body the wine will be. Full-bodied wines are known for being heavier, robust, and more complex. Light-bodied wines will be much lighter on the tongue, while medium-bodied falls somewhere in between. Red and white wines fall all over the spectrum in regards to full or light bodied. Although it may seem red wines are typically more full-bodied keep in mind that quality Chardonnay and Sauvignon Blanc, some of the most produced white-wines, are considered full-body. Wine fermented or matured in oak will likely give wine more body and weight. A common misconception is the higher-bodied the wine, the better quality. However, balance ultimately determines quality.

7. Brix

A brix is a scale used to measure sugar in unfermented grapes. Sugar levels are important to monitor because they are directly related to potential alcohol content and give insight on whether the grapes are ready to be harvested. A refractometer is the device used for measuring sugar content. To use a refractometer, a grape juice sample is placed onto the lense. The refractometer works similar to a prism and measures soluble solids. The device is able to determine the sugar content by observing the reflection of light. Each gram of sugar fermented will turn into half a gram of alcohol. Winemakers usually look for a brix level between 19.5 and up to 26 depending on the variety being produced. Brix is the main factor in final alcohol content but the type of grape, yeast strain, and fermentation methods also play a role.

8. Cap

The cap is referred to during fermentation. Grape solids such as pits, skins, and stems will rise to the top of the container during fermentation. The cap is a big factor in determining wine color, tannins, and weight. When the cap forms, many will "punch down" the cap to help the solids integrate into the container. In the beginning, punching down aids in bringing oxygen to the yeast, which allows them to get going. Punching down during the whole process helps the wine develop a richer, more complex structure and movement prevents mold from growing in the cap.

9. Color

The color of a wine is more important than you may think. An experienced wine connoisseur will look at color and have insight to the wine's age and quality. The pigment for red wine is called anthocyanin which comes from the skin of the grape. As anthocyanin is released during fermentation, the wine is stained. Young red wines will have a red, violet, or blue tint. The acidity can also be assessed by looking at the hue of a wine. Red wines with a strong red hue typically have higher acidity while blueish tints have lower, and violet hues in the middle. It is also true that lighter bodied wines will not have the same deep pigments as a bolder wine. As white wines age, they grow darker in color.

10. Dry, Sweet, & Semi

The terms listed above take into account the overall residual sugar left in a wine. Dry wines are the least sweet while sweet wines contain more sugar. There are many different "semi's" in between that all reference the sweetness of a wine. A dry wine will mean that all grape sugars were converted to alcohol during fermentation while a semi wine will still have sugar left over from fermentation. Some may perceive sweetness in dry wines due to the ripeness of the grapes when harvested or the bouquet developed.

11. Enology

The science of wine production. A professional winemaker is called an enologist. A wine connoisseur can be called an enophile.

12. Fermentation

Fermentation is a process in winemaking that uses yeast to convert sugars into alcohol. The process starts during crushing and lasts until after bottling. When the grapes are crushed, the natural yeast in the grapes are exposed to the natural sugars for the first time which causes fermentation. Fermentation is the process that makes wine alcoholic. The two stages of fermentation are the primary and secondary. The primary stage lasts up to a week and the secondary lasts up to two weeks. In the first stage, natural yeasts are busy eating sugar. Cultured yeast is added to help control the process. The second stage is much calmer and slower. By the second stage, most of the sugar has been consumed and the yeast are fighting to find food. In both stages the wine is sitting in barrels, oak, or a bottle to prolong fermentation. Temperature, speed, and oxygen levels are heavily monitored. Higher temperatures are used in red wines to bring out complexity, while white wines fermented at cooler temperatures to bring out fruit flavors. Malolactic fermentation and carbonic maceration are two other types of fermentation that will be discussed later in the chapter.

13. Finish

The finish of a wine is the impression left in the mouth after the wine is swallowed. The finish of a wine also takes into account how long the flavors of the wine linger and the textural impact in regards to a dry or crisp impact. This range can last for seconds or even minutes. Many believe that a long finish is a sign of quality. Most quality, great wines will have a long, complex finish. Terms such as moderate length, find length, good length, modest length, lengthy finish, and ample length are used when claiming a finish. Lingering flavor can be described as persistent, lasting, lingering, crescendo, and gaining momentum.

14. Grafting

Grafting is a helpful technique that allows the scion (top part) and rootstock (roots) to be blended. This is done by splicing the top of one variety onto the bottom of another. Grafting allows a variety to keep desirable characteristics but uses a sturdier root to offset climate conditions. Often a rootstock will be alive, and a scion will be added. Other Times, varieties are grafted before being planted.

15. Legs

Legs appear after a wine glass has been swirled, and refers to the streaks lingering on the side of the glass. The droplets or tears left behind will give indication to the wine's alcohol content. It's suggested that the more legs that appear, the higher the alcohol content. Sweeter wines or more viscous wines will have slower tears when falling down the glass. The legs of a wine won't indicate the level of quality but rather may give insight before consuming. The correct method is to hold the glass at an angle and allow the wine to gather on one side of the glass. Next, return the glass to an upright position and observe the legs to gather information on viscosity. A lot of legs means the alcohol has a higher alcohol content. When swirling the glass, the same can be noted and aromas can be observed due to the evaporation occurring.

16. Maceration

Maceration is the process of grape juice and skins fermenting together. The skins must be included to be considered maceration. The contact of the skin and juice contributes to color, tannins, and aroma in red wine. This can be compared to steeping tea. If the tea bag is the grape skin and the water is the juice, think of the effect the tea bag has on the water the longer the two sit. The longer the tea bag sits in the water, the more concentrated, full flavor the water gains. Without maceration, red wine would not have its red color.

17. Malolactic Fermentation

Malolactic fermentation is the act of turning malic acid into lactic acid. The reason malolactic acid is not desirable is because the wine then has potential to ferment in the bottle. Malolactic acid is unstable and can trigger carbon dioxide in the bottle, causing the wine to spoil. Malolactic fermentation is different from primary fermentation because it involves bacteria fermenting. Most red wines will go through malolactic fermentation, with some white wines (not all) as well. Malolactic Fermentation softens acidity, increases pH, adds new aromas, and stabilizes microbials. Some vintners may opt to skip malolactic fermentation which results in higher acidity, preservation of aromatic compounds, and filtration. Malic acid can be compared to a granny smith, while lactic acid gives a softer taste comparable to yogurt. In white wines, a compound called diacetyl will form during malolactic fermentation and is comparable to buttered popcorn. There are a few factors that will disallow a wine to go through malolactic fermentation. If a wine is over 14.5 in alcohol content, has an initial pH below 3.2, or has a temperature below 55 degrees Fahrenheit malolactic fermentation may not be possible. Vintners who do not wish to go through malolactic fermentation can expose the wine to sulfur dioxide or lower the temperature to below 50 degrees Fahrenheit to inhibit the process.

18. Organic

Organic wine is the act of growing grapes and producing wine without pesticides, synthetic fertilizers, genetic engineering, chemical additives, ionizing radiation, or sewage sludge. Pests and soil are managed using various, natural methods. Organic wine in the United States is either made with organically grown grapes or just labeled "organic." The difference between the two is whether sulfur dioxide is added or not. Wines that claim organically grown grapes may have sulfites added for preservation purposes whereas organic wines may only have a small amount of naturally occurring sulfites. The USDA organic label is only found on organic wines.

19. pH

The scale used to measure the acidity and hydrogen in wine. A higher pH or higher acid level is noticed within the first moments of drinking the wine. The pH level is important for an initial first impression in terms of the final product and can be described as a sharpness picked up by the sides of the palate. The pH level will change throughout the wine making process and is heavily monitored throughout. The pH level and acidity directly correlate and will be discussed throughout the book.

20. Pomace

Pomace or marc is the leftover solids after grapes have been pressed or fermented. Pomace from white wines are yielded after being pressed, and for red wine after fermentation. Pomace is made up of skins, seeds, and stems. Although pomace is a byproduct, pomace is still useful for most wineries. The remnants are typically rich in sugar, amino acids, nitrogen, and other nutrients. Pomace can be turned into compost and used in the winery as a natural fertilization. Pomace is also sold because it contains cream of tartar, an important ingredient in baking.

21. Press

Pressing is the process where juice is removed from grapes. This is typically done by a wine press, by hand, or using the weight of the grapes themself. Crushing typically takes place after crushing for white wines and after fermentation for red wines with some exceptions. Most vintners keep "free-run" juice and pressed wine separately. Pressed wine has a higher concentration than grape juice that didn't come from the press. Pressed juice will be bottled separately, divided, or added back into wine later on. Most wines are made from 85%-95% free-run juice and 10%-15% pressed juice (Margalit, 2003). Wine presses are generally categorized as batch or continuous with many different types of presses.

22. Pruning

Pruning is a method used to maintain the shape, vigor, and yield of a vine. Removing unnecessary branches, canes, and foliage ultimately helps the grapes receive the best conditions possible. Pruning keeps the vines healthy and productive. Pruning and the practices that come along with it decide grape yield, and components that go into the final taste. Pruning technically does give up crop, but is beneficial for quality purposes. When too many grapes are on the vine, nutrients and sunlight are shared. With fewer grapes, the vine puts more concentration into cultivating and maturing the attached grapes than worrying about the surplus growing. Pruning is essential a manipulation of the plant's energy. Pruning will allow grapes to reach optimal size and flavor which absolutely affects the quality of the wine. Pruning is unique to each vintner but typically happens during two phases: summer and winter. Winter pruning focuses on crowding and selecting growth points. Late winter and spring pruning is imperative for controlling foliage so the grapes receive enough sun, but not too much. Keeping the vines off the ground is also a goal during this time to prevent pest and disease.

23. Racking

Racking is the process of siphoning wine must from one container, and placing it in a new, clean container. The sole purpose of racking is to leave sediment behind. Racking aids in clarification and preventing rancid flavors. Racking is an important process that is performed about two to four times throughout the winemaking process. The first wine racking is done five to seven days into fermentation. This allows fermentation to continue and introduces and air-lock. Air-locks protect the wine from contaminants. The first racking will likely remove about seventy to eighty percent of sediment. The other twenty to thirty percent of sediment will take much longer to build up, thus having to rack a few more times. The second racking occurs when fermentation activity is complete. The third racking occurs near the end of the process to rid any leftover sediment. Other raracking should take place when aging wines for a long period of time. However, too many rackings can disrupt the wine's natural processes. More on this will be discussed later in the book.

24. Tannins

Tannin is the phrase that takes into account several compounds such as flavonoids, flavonols, catechins, anthocyanins, and ellagitannins. These compounds are naturally occuring in grapes and are mostly found in the skin, seeds, and stems. Tannin is found frequently in nature and is not specific to grapes. Apple skins, walnut skins, intense dark chocolate, green tea, plants, and tree bark are some examples. Tannin's purpose to a plant is to fight infection, pests, fungal diseases, and UV rays from the sun. Tannin's purpose in wine gives pigment, bitterness, and a mouth-drying sensation. Many tannins have been found to be healthy for the body fighting cholesterol and obesity (Tebib, Besançon, & Rouanet, 1994).

25. Terroir

Terroir is the term used that takes into account a region's climate, soil, and terrain with consideration to the wine's final taste. Terroir is the effect on wine from the listed elements. Climate is one of the elements to terroir that can be broken into two climates, cool and warm. Warm climates have a reputation of higher alcohol wines due to an abundance or sugar, while cool climates have a reputation for higher acidity, lower sugar wines. Soil affects terroir similarly to water passing over a teabag. Soil does leave an impression on the final produce. The third element is terrain. Geological features such as mountains, valleys, and water source play into terrain along with other plants, microbes, and trees. Terrain takes into account the effect elevation has on wine as well. Lastly, tradition can also affect a region's terroir. In well-known, wine growing regions, some of the practices and techniques will be included when the term terroir is used.

26. Sommelier

A sommelier is a wine professional who has undergone formal training. Duties include everything to do with wine service, wine and food pairings, and wine storage. A sommelier is often referred to as a steward, and someone who works at a fine dining establishment. A modern sommelier may be tasked with developing a wine list, training staff, and working alongside a chef for more innovative food pairings. There are four levels a sommelier may obtain: sommelier into, certified sommelier, advanced sommelier, and master sommelier. A master sommelier must receive an invitation for this certification and there are only 236 master sommeliers worldwide, 149 in the United States (Lee-Sedgwick, 2018).

27. Vintage

The term vintage refers to the year the grapes were harvested. A wine bottle that claims vintage on the label should be accompanied by a number, the year the grapes were harvested. Vintage is not always a determinant to the wine's quality. For example, champagne is often classified as vintage or non vintage. A non vintage wine will contain grapes from different years which is seen as an advantage to some varieties.

28. Vintner

The proper term used to refer to a winemaker or someone engaged with the wine making process.

29. Viticulture

Viticulture is the science, production, and study of grapes. When grapes are being grown for wine production, viniculture is another term used. Vitis vinifera is the most commonly cultivated grape for wine production.

30. Yeast

Yeast is the main factor that differentiates grape juice and wine. When no oxygen is present, yeast converts sugar into alcohol and carbon dioxide. This process is called fermentation and is crucial for a winemaker. The most common yeast pertaining to winemaking is Saccharomyces cerevisiae. This yeast is easily manipulated and gets the job done. Kloeckera and Candida genera are natural yeasts already found in grapes, and begin the fermentation process naturally, after the grapes are crushed.

All About Serving and Tasting Wine

Now that you know the basic terminology, it's important to learn how to properly taste wine. Knowing how to properly taste wine is not a novelty, and will help you produce better wine. Tasting is one of the many ways vintners make important wine decisions. After producing a variety, the next variety produced is likely adjusted. This section will teach you how to properly sample your wine, so you can let customers know everything they should experience. Be sure to serve or consume wine at the best temperature for the variety. Most red wines are optimal between fifty-five and sixty-five degrees Fahrenheit. White wines are optimal between forty-one degrees and forty-eight degrees Fahrenheit.

It can be argued that there are six main glasses of significant importance for drinking wine. Depending on the style and variety of wine, there is a glass best suited for optimal aromas, flavors, and notes. Full-bodied reds, light-bodied reds, rose or spicy reds, sparkling wine, light-bodied whites, and fortified sweets arguably all have their own glass for consumption. There are many scientific studies surrounding the importance of drinking a certain type of wine from a certain type of glass. It's been shown that the glass in which the wine is consumed from affects the density and position of vapors through the opening of a glass. It is also thought that stem or stemless is less of a factor but may impact the temperature of the wine since fingers transfer heat into the glass. Below you will find an image showing four examples of the different styles of glasses available. The furthest left shows a flute style glass perfect for sparkling wines. The narrow, vertical style encourages the bouquet from sparkling wines to bubble to the top. The second from the left is perfect for red wines. Notice the fishbowl like style, which enhances the acidity and intensity found in full-bodied red wines. This glass encourages the drinker to take in all of the aromas. The third glass from the left is suitable for white wines. The style pairs nicely with crisp, easy to drink wine. The last glass shown is quite universal and encourages any wine to breathe and flow onto the tongue smoothly.

Step One: The first step to experiencing the aromas of the wine, is to gently swirl the glass. The wine should touch the sides of the glass creating a larger surface area for sampling. The swirling increases the wine's contact with the air and intensifies the aromas. You can swirl the wine by holding the glass or by the base of the stem. Smell your wine as you swirl and notice aromas such as fruits, spices, herbs, and flowers. Be sure to repeat this step several times, swirling before each smell. This is an important step and shouldn't be skipped. The overall taste of a wine encompasses smell and taste.

Step Two: After swirling, it's time to taste the wine. Take small sips. Roll the wine across your tongue stimulating your taste buds. Swish the wine around your mouth for five to ten seconds. Return the wine to the center of the tongue before swallowing. After swallowing notice how long the sensations last in your mouth. This will tell you about the aftertaste and finish of the wine. The finish will tell you about the quality of the wine. The longer, more complex the finish the higher quality the wine is to be considered.

What you Should Know: Competition, Timeline, &

Costs

The wine industry is booming with endless possibilities for the consumer. As a businessman or businesswoman, it's important to recognize the demand for wine while taking into consideration the different perspectives the consumer may have. When stepping into the wine section at a grocery store, hundreds of bottles are in view. However, don't let this intimidate you. A successful wine business is still very possible in this day and age. The consumer is always looking to find new niche wines that stand out and local wineries can spark huge intrigue. First impressions are everything in the wine world so it's important to understand the art thoroughly. Luckily, my expertise will be able to guide you and get your business where you want it to be.

When starting a wine business, it's important to understand the timeline. Most winemakers will tell you their wine took years to perfect, and that success didn't come overnight. A successful wine business will require determination, investments, and a lot of hard work. The wine industry is a tough, competitive, and expensive business. Accepting those three factors from the get-go is imperative for long-term success. The timeline from preparing the site to actually bottling wine can take years. Prior to planting any vines, the preparation of a plot can take up to two years. Finding a suitable plot may take awhile. Before anything is planted the soil needs to be in proper condition. The vineyard needs to have a solid design and rootstocks or type of vines must be decided. Materials, equipment, possible fencing, and a strong business plan should be in place before planting. When all of the details are sound, it will take additional time for the grapes to mature even after a few harvests. The first vintage may not be even bottled for another two years after the first harvest. Being patient for your debut will allow time to perfect your craft and win over customers with a positive first impression.

After understanding the wine industry and timeline, the other thing to note are the costs. The costs will range depending on location and size, but be aware that a wine business is a huge investment. It's estimated that starting a very small, backyard vineyard costs anywhere between $35,000 and $45,000 per acre (8 Steps to Owning Your Own Vineyard, 2018). After development annual costs are estimated from $15,000 to $20,000 per acre in the first three years, just to keep the vines alive (Goldstein, 2020). Other purchases to keep in mind are equipment (refrigeration, cellar equipment, winery buildings, farming trucks), vines, fermentation and storage processes, bottling line, payroll, insurance, office space, and possibly a tasting room. According to a study by Washington State

University, operating a winery costs between $600,000 and $2.3 million depending on the amount of cases produced. Because each situation will be different, keep in mind that some of the numbers may be lower or higher depending on your situation. However, be aware that having a business and financial plan mapped out right away is imperative.

Creating a Business Plan

In this section, we'll go over some of the costs and planning required for a startup vineyard business. Having a sound business before getting started will help you prepare for what's ahead and limit surprises. Starting a vineyard like many other businesses is always an investment that has risks. Acknowledging as many risks as possible and finding a solution beforehand will help your business immensely so time can be focused elsewhere. Remember that this section is a basic guideline and many of these factors will depend on the area you're in. Use this section as a template and fill in more specific numbers by conducting your own research of the area.

The first thing to consider to start a vineyard is land. Afterall, land is needed to grow grapes, the moneymakers. A significant amount of land is needed to start looking at what's available to get an idea of the initial costs. The first year's budget should be maximized greatly per acre in comparison to the following years. An example from TRUiC (2019), an information company, if you budget $12,000 an acre the first year, the second should drop down in costs to $1,400 per acre, with year three even less costing $1,000 an acre. These estimates do not take into account the rent or costs associated with owning or leasing the land, so factor those in. The land will likely cost the most, as it should if you're investing in quality. In prime real estate in California an acre may cost anywhere from $400,000 to $500,000 per acre (Thomson, 2019). Other states such as Virginia have much lower price per acre costs ranging from $11,000 to $30,000 (Franson, 2012).

The next part of your business plan should include ongoing expenses. In addition ownership/lease expenses, you need to start thinking about the maintenance required to actually grow the plants. Labor, insurance, irrigation, vine management, maintenance, and machine repairs should be documented in the budget. Plan for $8,000 total per acre for the first three years (Thomson, 2012). When the vines start to produce profitable grapes, this cost should steady at about $1,500 to $2,000 an acre after the first years.

Make a plan to estimate profit. A vineyard in full production is estimated to make an annual return of $2,500 to $5,000 per acre although these numbers depend on quality and demand (Thomson, 2012). For example, a 35-acre vineyard that earns $2,500 annually per acre can expect $88,000 profit.

Social media and marketing is the next area to focus your budget. Social media strategy will be discussed in a later chapter, but absolutely include costs for this in your initial business plan. Knowing how to get people excited and when, is key for creating loyal customers. Getting your name out there, especially in high competition areas, will set you apart from other wineries. It's suggested to allocate 10% to 13% of your total budget to marketing in the beginning (Frazier, 2017). This money set aside will be useful for website costs, marketing campaigns, and hired/professional help. Use this part of your business plan to establish your target audience.

Let's say, for example, that you have a perfectly produced, beautiful bottle ready to go. Who will you sell it to? Envision your customers based on the area you live. Start brainstorming ideas for label design and what you will name your business.

After you've had a good start to your business plan, focus on establishing your business as a legal entity. An LLC (Limited Liability Company) is recommended to prevent you from being personally liable if the business is sued. If you don't want to establish the business as an LLC, consider a corporation or DBA (Doing Business As). Enroll a registered agent if you're unfamiliar. Before opening your business, you'll need to register for many different state and federal taxes. Apply for an EIN (Employer Identification Number) with the IRS through their website, fax, or mail. It's suggested to open a business bank account and apply for a business credit card. This will protect personal assets, keep financial spending organized, build credit, and will make filing taxes easier. Remember to keep a log of all spending and seek help from an accountant if necessary.

Before opening your business, learn what types of permits and licenses are needed. Document them all in your master business plan and check them off as you go. This will be discussed in greater detail later in the book but check with your town, city, or county clerk's office to be thorough. Know that federal, state, and local licensing is likely required. Start looking into insurance policies for your business especially when working with employees to receive aid in any worker's compensation claims.

After you feel you have a great business plan, give the business plan to an accountant and a lawyer to look everything over. Although this may be an extra expense, having professionals help will put you on the right track and eliminate potential legal worries. The following page will give a helpful checklist as a guideline to get your business plan started.

Business Plan Guide Checklist:

1. Make a detailed list of costs expected costs for the next 3 years, as it may take that long to see a profit. Take into consideration land, equipment, plants, tools, pesticides, irrigation system, trellises, maintenance, salaries, marketing, licenses, insurance, repairs, etc. Hire an accountant if necessary.
2. Make a plan for expected revenue for the next 3 years.
3. Form a legal entity by establishing an LLC, DBA, or Corporation.
4. Apply for an EIN through the IRS. Register for state and federal taxes.
5. Open a business bank account and obtain a business credit card.
6. Obtain permits and licenses on the local, state, and federal level. Visit a local clerk's office for more information for the requirements of your area. Receive a certificate of occupancy and familiarize yourself with liquor licensing. Consult with a lawyer if necessary.
7. Get an insurance policy for your business.
8. Brainstorm on what your brand's mission is and target audience. Make a plan for social media presence.

Chapter 2: Getting Started

Choosing and understanding your location is the most important factor in starting a wine business. Because microbiomes and different climates exist all over the world, it's important to know your advantages and disadvantages from the start. This chapter will explain important elements to take into consideration when choosing a plot. The second part of the chapter will teach you the basics on setting up your vineyard. Spacing, orientation, trellis information, and overall design will be discussed.

Components for Selecting a Site

Almost every decision made during the life of a vineyard will be in consideration of site selection. The character of your land needs to be evaluated to maximize profit. Certain wines prosper in certain areas solely due to the components of the site. This section will discuss elevation, slope, aspect/positioning, and why the history of the site matters. Knowing what you're working with will lead to the best business decisions possible.

The first element to discuss for a site is **elevation**. Sea level and elevation variations in your site should be studied. Absolute elevation refers to feet above sea level. Usually, high latitude locations seek lower elevations, while low latitude locations seek higher elevations. This is due to the possibility of frost which occurs at higher elevations. Frost shortens the growing season for the vines and unpredictable frost can set you back.

Knowing the highest point of your plot is important for relative elevation. Relative elevation takes into account the highest and lowest parts of a plot. Even at a favorable absolute elevation, a vineyard still may be susceptible to spring and fall frosts due to its location. In a mountainous region, it's best for a vineyard to be located where freeze and frost temperatures are less likely. This is referred to as the thermal belt. Planting at the lowest elevation of a plot may make it susceptible to cold air. Cold air is heavier than warm air, and settles in lower elevations. In the event of a frost or freeze, you'll want sufficient drainage. Better air and better drainage will occur at the highest point of a plot. Standing water or flooding is not good for the vines long term as the oxygen the root systems receive will be limited.

Both absolute and relative elevation play a role in the finished product. Wines produced in higher altitudes have an advantage over those produced from a valley floor. Vines grown on the mountainside receive more direct/concentrated sunlight, large temperature changes, and quality drainage. Although less grapes are typically produced in comparison to those coming from a valley floor, the grapes that survive all of these elements are unique. The berries that survive are full of character and pack a greater punch. Higher elevations and more concentrated sunlight allow the grapes to endure deeper pigment concentration. This is similar to getting a tan, while baking in the sun. The sun helps the grapes to form a tough skin which is essential for making aged wine.

On the mountainside, temperatures then plumet at night making sugar production stop. The intense balance of hot days and cool nights forces the grape to ripen slowly. Flavors develop on a more mature level in this process. In a valley, temperatures usually stay warm producing a different tasting result. The drainage system is also different for mountainside vines and valley vines. The rain that occurs on the mountain sweeps nutrients found in soil down to the base.

With a lack of superficial nutrients, the vine is forced to grow deep into the earth, creating a more mature and complex rooting system. In a valley, the soil is typically much more fertile and vine roots may only grow a few feet deep into the soil. Although valley production may be higher, yielding more grapes, mountainside production yields fewer, but substantially higher quality grapes that have much more character. Whether you're on a mountainside or in a valley, knowing the difference will help create the best strategy for business.

The next element to take into consideration is **latitude**. Like real estate, location is everything. The location of a vineyard is one of the main factors in the character the wine holds. Typically vineyards require a temperate climate which occurs in the middle latitudes. Temperate means moderate, and these climates don't have extreme variations in temperature or precipitation. These regions typically have four seasons as well. The best grape growing zones are said to be between the latitudes of 30 and 50 degrees north of the equator and 30 to 50 degrees south of the equator. This is known as the wine belt. Places closer to the equator usually bare too high of a temperature. Although latitude plays a huge role in winemaking, not all areas that fall in between 30 and 50 degrees latitude are suitable. For example, in some of these locations you will find scenarios such as lakes, dry areas, cities, and swamps which are not suitable for winemaking. Popular wine regions in France, California, Australia, Argentina, and South Africa all boast powerful wineries falling in the range given above. Italy, France, and Spain produce most of the world's wine, producing a combined total of 122,800 hectoliters by volume (Top 15 Wine-Producing Countries, 2019). The United States was number four in world production yielding 23,600 hectoliters in regards to volume. Argentina, Australia, Chile, and South Africa were to follow at lesser amounts. When looking at a map, you will find that all of the countries mentioned fall in the latitude range given.

Furthermore within those belts, you will find that some vines grow best in cooler conditions whereas others in warmer conditions. Knowing the latitude location of your plot is helpful to deciding which grapes will prosper best for you. (This topic will be discussed further in Chapter 3.) Although most wine production occurs in the wine belt, there are exceptions. Although conditions may not be completely suitable, winemakers are trying their hand in other regions. Brazil, Great Britain, Norway,

Denmark and Thailand are a few locations producing wineries despite not being in prime location.

The third aspect in picking a location for winemaking is **slope**. The slope of a plot takes the degree of inclination and expresses the information as a percentage. A flat vineyard would have a zero percent slope, while a vertical cliff will have a one hundred percent slope. Slope is important for a vineyard because slope is related to air flow, soil drainage, water drainage, temperature flow, and erosion. The slope at a particular vineyard will rely on other factors. First, the further north or south from the equator, the steeper the slope needs to be to receive optimal sunshine and heating. It's recommended that a vineyard should have a slope of at least five to ten percent. Other vineyards may have a higher slope, but the higher the percent the more difficult the vineyard is to manage. Flat vineyards are easy to manage and have less erosion problems, but are up against cold air inversion in locations with frost. A vineyard with a small slope, but that isn't flat will have better air flow but erosion then starts to become a problem. As the slope increases, better air flow and water drainage occurs, but erosion, row orientation, and management becomes difficult. As you can see, the slope of a vineyard has positive and negative attributes. Gentle slopes over flatter terrain are usually prefered for easier management and some of the benefits a slope offers.

The last component to consider when choosing a location is **aspect**. Aspect refers to the direction the vineyard's slope faces - north, east, south, or west. Aspect determines the total heat balance of a vineyard and the angle that the sun hits the vines. Aspect is most important the further you get away from the equator in the wine belt, northern hemisphere or southern hemisphere. These places receive less radiation from the sun. Because the sun plays an imperative role in the development and ripening of the grape, aspect is always important. In the northern hemisphere, cooler regions benefit most from south-west facing slopes to receive the most sunlight hours as possible. Southern facing slopes warm earlier in the spring and the vines are able to break earlier. An earlier break translates into an earlier bloom and harvest as long as no unpredictable frost occurs.

In the Northern Hemisphere, warmer climates benefit most from northern facing slopes to allow the grapes to mature slowly, without the constant rays from a more intense sun. Western aspects are sometimes used for grapes that mature later in the season. An example is Cabernet Sauvignon which prefers waning heat and fall sunshine. Lastly, eastern aspects are also beneficial to certain varieties because they receive morning sun quickly which allows dew to dry quickly. An eastern aspect lowers the risk for rot and disease amount the grapes. In the Southern Hemisphere, these orientations are reversed because of their position with the sun, but the same concepts prevail.

Once a plot has been decided on, designing the vineyard is next. Deciding row orientation, row spacing, vine spacing, and trellis design are the next steps. These three elements will correlate with the quantity and quality of production. In regards to **row orientation**, vineyards usually are planted from north to south to ensure even, maximum sun exposure throughout the entire vineyard. Remember to keep your chosen aspect in mind when deciding row orientation. However, there are situations where other orientations may be necessary.

For example, if the plot is rectangular it's best to plant fewer, longer rows instead of multiple, short rows. Short rows on a rectangular property will create more of a hassle when doing maintenance on the grapes and working with machinery. Limiting the amount of end posts on a site, and maximizing the length of a row will cut down on costs. Typical row lengths are between 200 and 300 feet, with some breaks in between (Stafne, 2019) for convenience. Erosion should also be considered when deciding row orientation. If the plot has a steep slope, it's best to work horizontally with the land instead of planting rows up and down (vertically) which could cause more erosion and more work. Many vineyards will base row orientation off of the erosion potential the plot has.

To maximize quality and quantity, **row spacing** should be given a lot of thought. Row spacing should be decided before buying any equipment, or if you own equipment already, the equipment owned should decide row spacing. Row width should be determined based on the height of the canopy or overall plant/vine. The vines are unable to receive critical sunlight when they're all growing on top of each other. As a rule of thumb, the ratio between canopy height and row width should be one to one. For example, if the canopy height is planted for six feet, the row width should be at least six feet or more. Not only is vertical height a factor, but the length horizontally or vine vigor should also be considered. Vine vigor refers to the growth rate of a vine and it's shoots. Soil, fertilizer, weather, bugs, and pests can all be factors. (More information on vine vigor will be discussed in Chapter Three.) After deciding which types of grapes will be grown on the property, the business and plot plan should be revisited to ensure the correct factors have been chosen. Some species of grapes have low vigor while some have high. Knowing the specifics of the grape and how they grow will help determine row spacing.

In regards to **vine spacing**, the area between vines usually ranges from three feet to twelve feet although six to eight feet is the most common (Stafne, 2019). The first element to consider when spacing vines is soil potential. Soils containing a lot of nutrients and water typically have a higher capacity and can allow the vines to be placed further apart. Low nutrient soils that hold less water require the vines to be closer together to maximize yield and quality. As a rule of thumb, high potential soils will yield larger vines while low potential soils yield smaller vines. (Soil is discussed further later in the chapter.)

Knowing the type of soil you're working with will help decide the first component in vine spacing. The second concept to consider is that vines placed close together will be in competition for nutrients. This is not always a bad thing because it's always "survival of the fittest," and higher quality grapes are produced. When vines are placed close together, the density of the crop or canopy is more ideal in terms of quantity. However, keep in mind that too dense of a canopy will lead to unnecessary shading and too much crowding. In summary, the grape variety chosen and the potential of the soil should dictate the spacing between vines. Start in the general range of six to eight feet and adjust to the rigorousness of the grape variety.

When choosing a site, learn about the soil right away. **Soil** is important for providing anchorage, water, and nutrients. Topsoil and subsoil are typically different and should be assessed separately. Topsoil is responsible for supporting the root system and food network. On some occasions, thin topsoil may need nutrient additions depending on the variety being grown. This is common for champagne regions, for example. Subsoil is responsible for drainage, and a deeper root system.

The first element to good soil is sufficient drainage. All plants need water to survive, similar to humans. The best soil for wine will drain into deeper parts of the topography which in turn will create a deeper rooting system. The soil should be clear from obstacles that will hinder the vine from growing deep into the topography. The deeper, more complex the root system, the higher quality wine is. A simple explanation is that topsoil water has less minerals and nutrients than that found deep into the earth. The water the vine receives is linked to the overall health and complexity of the vine. If natural drainage is not happening on your site, a drainage system should be installed. Poor drainage results in restricted vine growth, and restirected vine size. Tiling is the practice of removing excess surface water. Perforated plastic pipes are used to form a network, and installed under the vineyard. Subsequently, when the water table rises the pipes work to mitigate the water to a drainage ditch. The tile/pipe networks aids in moving the water away from the vineyard.

The second element to soil is fertility and nutrients. It's generally thought that grape varieties do best in low nutrient, poor soil. However, there are still compounds in the vines. In many cases, organic matter is added to the soil annually. The most important nutrient to a vine is nitrogen. Nitrogen is responsible for all green matter produced on the vine. Too much nitrogen will trigger overgrowth which can be a problem. So although nitrogen is important, too much nitrogen can be detrimental if a vine is not carefully managed.

Phosphorus, potassium, trace magnesium, trace iron, and trace zinc are other beneficial nutrients. Areas that have high nutrient soils need careful management to control the vigor of the vine. The pH of a soil should also be examined. Low pH, or high acid soil will produce grapes lower in acidity while high pH, lower acid soil will produce grapes with higher acidity. As you know, the level of acid in grapes is crucial for harvest time and overall production. All and all, a well balanced soil will produce the best quality fruit.

The most popular soil types in wine growing regions are chalk, clay, gravel, granite, limestone, sand, and slate. Chalk is a type of limestone that is very porous. Chalk conducts good drainage but balances this with good water retention. This combination is great for the wines to receive what they need. An area that proves the success of chalk soil is the Champagne region of France. Chalk soil can be beneficial in hot, arid regions for its ability to absorb and store water for future times.

The second type of soil to discuss is clay. Clay is known for its poor drainage but makes up for this quality by having efficient water retention. Clay is often used as a subsoil for this reason. Clay soils may not be suitable for cold areas as they take longer to heat up after winter, which can delay a growing season. Clay soil is found and valued in Bordeaux, France for producing wines with lots of structure and color.

The third soil to consider is gravel. Because gravel is not able to retain water well, gravel supports efficient drainage. Gravel may be helpful for regions that experience high amounts of rainfall and regions with cold nights. Gravel absorbs heat during the day and acts as a buffer overnight in cold regions. Grapes that have a late harvest and ripen late, like Cabernet Sauvignon can benefit from gravel soils.

Granite, the fourth type of soil, is made of hard, crystalline minerals. Granite heats faster than other soils and also retains the heat. Granite soils have low fertility and conduct efficient drainage. Granite soils promote higher acidity wines and deep rooted vines. Riesling, Syrah, and Zinfandel are examples of varieties that develop well in granite heavy soils.

Limestone is the fifth soil to mention and is thought by some to be the finest wine-producing soil of them all. Limestone is made mostly of calcium carbonate and is derived from prehistoric, decomposed sea life and reefs. Limestone is able to provide efficient drainage and retain water which is helpful for dryer periods. Limestone is a rarer soil type but is found in some regions of California. Fine Chardonnay and Pinot Noir are two types that thrive in Limestone soils.

Sandy soil contains at least half sand. Sand is often mixed into other soils for added benefits. Sandy soils are coarse, porous, and drain water the best out of other soils. Because sand drains water so well, sandy soil is extremely low in nutrients. Desert regions are less likely to benefit from this because such a region needs water retention in soil. Areas with heavy rains are the most likely to benefit from sandy soil. Another positive attribute of sandy soil is the diminished risk of disease and Phylloxera (a detrimental pest).

Tempranillo and styles of Zinfandel or Cabernet Sauvignon are examples of varieties grown in sandy soils.

The last soil to mention is slate. Slate is composed of clay, shale, and other compounds. Slate is a hard, dark, rock-like slab. Slate is able to hold moisture, warm quickly, and retain heat. Slate is beneficial during cooler nights because it has the capacity to insulate the vine. Slate is a soil that helps grapes ripen and is durable under harsh conditions unlike other soils. Germany is known for its award winning Rieslings which are mainly grown in slate heavy soils.

Although you may not have perfect conditions across the board, choosing a grape that does not do well in a specific soil will hinder you from expanding revenue. In some cases it's like trying to grow pineapples in Antarctica. When choosing a site, learn about the soil promptly. Soil shouldn't hinder you from growing many varieties but may require the addition of other soils in the plot. In many instances vineyards will play off the original soil but manipulate or mix the topsoil for better results.

The Design and Trellis System

After deciding upon row orientation, row spacing, and vine spacing, a **trellis system** should be chosen.

Trellises are the bones of a vineyard. A trellis is responsible for supporting the vine and training vines to grow in a position that's most suitable for sunlight and maintenance. A trellis is usually made of wood or metal and is drilled into the ground, similar to a fence. The trellis will support the weight of the vines and spread the canopy out evenly. A vine that is too crowded is susceptible to mildew, rot, disease, and shade.

Some of the most popular trellis designs are named high cordon, umbrella kniffen, and vertical shoot position (VPS). The simplest trellis design, high cordon, is useful for American and American-french grape varieties. A high cordon trellis uses one wire and stretches it between posts. The arms or cordon of the grapevine grow up, then horizontally along the wire. The umbrella kniffen system is also suitable for American grapes. This system uses two wires, upper and lower, and stretches them between posts. The vines grow to the upper wire, then are typically tied to the lower wire. French wines usually favor the VPS system due to their upward growth tendencies. In a VPS system, four to six wires are stretched between posts. The vine is trained to grow on the bottom posts, while the shoots grow upwards.

Choosing a trellis design should be decided on the type of grapes being grown, their size, and how much the budget allows. Trellis systems can be complex and expensive to design. Keep in mind the main goal is to maximize sunlight and support the vines when installing a trellis system.

Chapter 3: All About Grapes

The following chapter will take you through everything you need to know about grapes. Starting with wine regions of the USA, this section will teach you about the various regions growing grapes and let you into some of their success. This section will be helpful to you, as it is region specific and will give tips pertaining to your area in the United States. You will learn that different varieties thrive in different regions. The variety you decide to grow should be inspired by science or local success. Knowing you're growing a grape successful to your area will allow you to focus on putting your spin on the final product!

The second half of this chapter will discuss selecting grapes and what to look out for. Brief introductions will be given to the most commonly known grapes, and the varieties that do best in the United States. Knowing about different grapes will help you perfect your practice and may come in handy if you chose to blend your wine. This section will also explain the differences between the grapes, and explain the climates in which they grow.

Wine Regions of the USA

In the United States specifically, vineyards are popping up almost in every state. While it's true that the US falls in the latitude most favorable for winemaking, how can such a large region be suitable for winemaking? The truth is, grapes can be grown in many regions under the right conditions due to the variations in their species. In the United States, California is likely the the state that comes to mind when considering wine production. But, you may be surprised to know that almost every state has at least one winery. Here is a breakdown of the top 10 states by amount of wineries according to Statista (2019):

1. California- 4,501
2. Oregon- 793
3. Washington- 792
4. New York- 403
5. Texas- 351
6. Virginia- 291
7. Pennsylvania- 285
8. Ohio- 254
9. Michigan- 191
10. North Carolina- 171

You may not be surprised that California boasts the most wineries. In fact if California was its own country, they would take fourth place in regards to the largest wine producers behind France, Italy, and Spain (Conway, 2019). You may also be surprised that each region of the United States is represented on the list. The Midwest, Northeast, Southeast, Southwest, and Western United States are ,remarkably, all represented. Even though a specific grape can be produced in two different regions, the climate will ultimately decide the final taste.

Cooler regions typically produce lower sugars, but higher acidity in grapes which leads to a more delicate taste. Warmer regions produce dense, bold wines with some exceptions of course. The most populous grapes grown in the United States are Chardonnay, Cabernet Sauvignon, Pinot Noir, Merlot, Zinfandel, Syrah, Pinot Gris, French Comombard, Sauvignon Blanc, and Rubired. Where the wine is produced will dictate the final taste.

Whether you're a fan of California produced wines or not, California does account for 90% of all American made wine (Wine Regions in California, 2018). In the West, California is the leading state for wine production. The climate and terrain of California is why California can produce so much wine, and quality wine at that. First, the rainfall in California sets the state apart from others in the United States. Rainfall in California really only happens for six months out of the year; primarily during the winter. The summers in California are warm and dry, which allows the grapes to ripen and develop bold flavors. Second, even though the sunlight is crucial for grape growing there is another element to California's success. The northern and central areas of California are located next to large bodies of water, the Pacific Ocean and San Francisco Bay. The winds that come off of the water provide proper cooling and occasional fog. That combination allows moisture in the air, and gives the grapes relief from the hot sun. Lastly, the mountains and hills of California produce the perfect soil for wine growing. The red and white clays of California are infused with a distinct mix of volcanic ash and minerals. Traces of Sandstone are also found in California soils. As you know, the majority of grapes thrive in drier soils. All of these unique elements together are not found in too many other places in the world, and are the reason California has become so successful.

California grows over 110 types of grapes, but Chardonnay takes the cake with over 95,000 acres planted (Guide to California Wines, 2019). Chardonnay is grown throughout the entire state of California, but is most successful in the Northern and Central Coasts. California produces a sweeter Chardonnay in comparison to other regions. Another notable white wine is White Zinfandel which is truly unique to California. Although grown in other regions of the world, the quality produced in California gained immense popularity. White Zinfandel carries a beautiful light pink color, although darker pinks are produced as well. Zinfandel grapes thrive in hot temperatures, and dry climates. The best California Zinfandel usually comes from higher elevations in the state which yield warm days and cooler nights.

In regards to red wines, California is known worldwide for their Cabernet Sauvignon, Pinot Noir, and Merlot. The Guide for California Wines (2019), states that Cabernet Sauvignon is grown across over 90,000 acres. Cabernet thrives in porous soil commonly found at the base of a mountain. Northern California, in places like Sonoma and Napa, the Cabernets produced are world-renowned for their exquisiteness. Northern California is also able to produce the tricky Pinot Noir, which requires a variety of factors like those found in California.

Moving into the Southwest, Texas is the main producer of wine in the region. Texas has eight areas in the state where wine is produced, but the Texas High Plains and Texas Hill Country are the most populous areas located in the central-western part of the state. Texas wine is sometimes compared to wines produced in Portugal, due to similar climate conditions. Texas is known for being dry, hot, and windy. Long days of sun are followed by cooler nights. The Ogallala Aquifer aids in water irrigation and gives tempering effects on high temperatures. The high winds in Texas are able to air out the vineyards and fight disease.

Cabernet Sauvignon and Tempranillo are the most commonly grown varieties. Tempranillo is a specialty of Texas and grows well. Tempranillo is considered a Spanish wine and is produced in other areas of the world like Spain, Portugal, Argentina, and Chile. Other varieties like Blanc du Bois and Black Spanish grapes are being planted in the southern part of Texas more and more due to their tolerance of humidity caused by the Gulf of Mexico. Wineries in Texas have a short growing season due to the intense temperature. Less time on the vine yields less ripe, softer flavors. Harvest is two months earlier than California, and three months earlier than regions in France, to combat the hot summer heat.

In the Southeastern region of the United States, Virginia and North Carolina make the list for states with the most wineries. In Virginia, wine has been in production since before colonial times. Many wineries were shut down during prohibition and later reopened. The hot, sticky climate of Virginia proves disastrous if attempting to grow the wrong grape variety. However, within Virginia lies many different microclimates and strong winds. The diverse geography and wind help Virginia to grow grapes that can fight off rot and prosper year to year. Chardonnay, Merlot, and Cabernet are grown inVirginia, but the best wines of the region are Rkatsiteli, Petit Manseng, Petit Verdot, and Norton. Virginia is known for growing Petit Manseng because

Virginia is the only place this grape is able to grow in the entire world.

In North Carolina, the eastern part of the state focuses mainly on Muscadine, a grape exclusively known to the state. Muscadine grapes thrive in hot, coastal climates and are even used for wines such as Scuppernong, Sangria, and Rosé varieties. In the western part of Virginia, Cabernet Sauvignon, Cabernet Franc, Viognier, and Merlot are produced.

The Northeast is led by New York and Penslylvannia in wine production. New York has 10 areas in the state where wine is produced, with the most notable being the Finger Lakes, Hudson Valley, Lake Erie Region, and Long Island. In New York many of these wine growing regions are near a water source. Because New York can reach high and low temperatures, the water bodies help warm the grapes in colder times, and cool the grapes in warmer times. Water acts as an insulator for most of the regions in New York. It's no secret that New York is one of the best states for growing grapes, even though not all are used for wine production. In addition to wine, New York produces large amounts of Concord and Catawba grapes which are heavily used for making jam, juice, and jelly. New York supplies the popular brand "Welch's" with most of their grapes with over 18,160 acres in the region (Welch's, 2020). Wine from New York is typically that of a French-American hybrids. Cabernet Franc is said to be the best in New York, second to Bordeaux, France. Another fine wine produced in New York is

Riesling. The Riesling produced in the Finger Lakes region has put New York on the map for it's quality. Riesling is easily grown in this region due to the unique layout of the glacier carved lakes, which insulate the grapes perfectly. Riesling produced in this area is typically dry and aromatic, setting it apart from sweeter versions produced elsewhere.

The climate in Pennsylvania is sometimes compared to northern France, northern Italy, Austria, and Germany. With mostly overcast skies and cool conditions, the grapes in this region usually have a lighter, more refreshing finish with white varieties being the most popular. Like New York, Riesling and Cabernet Franc are the best in the region.

In the Midwest, Ohio and Michigan are the top producers of the region. In both states, the grapes grown are chosen with cold temperatures in mind. In Ohio and Michigan, most vineyards are planted near a water source as water tends to be a good insulator. Sweeter wines are typically produced in Ohio with the most popular being Catawba, Niagara, Riesling, Pinot Gris, and Gewurztraminer. Although Ohio mainly produces sweet wine, other varieties like Pinot Noir and Cabernet Franc are also produced. In Michigan, "ice wine" is a specialty as the fruit is actually harvested while frozen. These grapes are harvested in December or January to make sure the grapes are fully frozen. Fruit wine, Riesling, Merlot, Pinot Grigio, and other hybrids are grown in Michigan as well.

In summary, no matter where you live in the United States a successful winery is possible. Different challenges will arise in each region, state, and territory within the stat based on the local geography. Luckily, in the next chapter we'll break it down into simple terms and help you choose a grape variety best suited for your situation.

Grape Selection

In this section, you'll learn all about the different varieties of grapes and how they perform in different scenarios. As you know, many factors go into the final taste of a wine. One variety produced in the west will likely taste different than the same variety produced in the east. Your location and factors like topography, temperatures, soil, and precipitation should be well documented so you can refer back to that information when selecting a grape variety. The variety you chose and its success will rely on the quality and quantity after harvest. Selecting the right variety for your area is likely the most important decision.

This section will go over the most popular grapes and the best grapes to grow in the United States. Grapes belong to the family Ampelidacae which includes a number of varieties. The genus Vitis refers to grape bearing vines. Vitis vinifera is the European and

Central Asian species grape vine and is responsible for producing almost all varieties of wine. Vitis vinifera boasts over four thousand different varieties and in terms of wine, tons of different flavors. Because Vitis vinifera has so many varieties, you're bound to find a type that will work well in your area.

Vitis vinifera has the ability to exchange genes and interbreed. Crossing is the term used that describes the outcome of fertilizing one variety with the pollen from a different variety. Crossing combines two varieties to produce another. Crossing or hybrids are illegal in the European Union for quality purposes, but legal in the United States.

You may be surprised to know that most of the world's wine use different varieties of the Vitis vinifera, but is attached to the roots of another species. This is the preferred method because most roots are native to the location and more successful at fighting off disease and pets. When European roots are exposed to American terrain, there are many new and unfamiliar factors that can lead to destruction. This was discovered and proven in the late 1880's when a pest called Phylloxera vastatrix devastated the wine industry worldwide. Because rootstock is primarily responsible for water and bringing up nutrients, the roots have little to no effect on the variety graphed on top.

The most common species of rootstocks used in the United States are Vitis riparia, Vitus rupestris, Vitis labrusca, Vitis berlandiera, and Vitis champini. Hybrids of the mentioned rootstocks are also available, and help perfect rootstock and climate success rates. For example, some of the hybrids will be more resistant to periods of drought, more suitable to calcareous soils, and more.

1. Cabernet Sauvignon

Cabernet Sauvignon is the offspring of two Vitis vinifera species: Cabernet Franc and Sauvignon Franc. Most Cabernets are characterized as being full bodied or medium bodied with a ABV (Alcohol by Volume) of over 13.5 percent and up to 15 percent. Cabernet berries are quite small in comparison to other grapes. The skin of the grape tough, thick, and deep purple which contributes to the end color; a deep ruby red.

Cabernet Sauvignon is widely produced but no two bottles will taste the same to an experienced taster. The final finish will depend on the region the Cab was produced. Methoxypyrazine is found in the variety and is responsible for giving notes of black pepper, bell pepper, black currant, bell peppers, and green peppercorn. While small amounts of these notes are favorable for the final taste of Cabernet Sauvignon, this is something to be managed, especially in warmer climates. High amounts of Methoxypyrazine will ruin the taste of Cabernet Sauvignon if not controlled, and leave the consumer with tastes of vegetables in their mouth. High amounts of Methoxypyrazine are mostly caused by too many green leaves hanging out on the vine which can be controlled by pruning. Spur pruning is recommended.

In cooler climates, Cabernets tend to be more herbaceous and give off mint, eucalyptus, leather, and earthy feels. In warmer and cooler regions, cabernet is typically aged in oak barrels which round out the flavors and add a hint of cedar, tobacco, smoke, vanilla, coconut, and spices. Cabernet is usually aged in oak barrels for one to two years or more to help soften the intense characteristics and tannins.

The best Cabs will do well in low nutrient, well drained soils. The combination will disallow multiple shoots and foliage growth. The energy is not focused on the entirety of the plant, but rather put into the production of the grapes in this scenario. The result is smaller, more concentrated berries. Cabernet requires a long growing season to allow flavors to develop and thrives in moderate to warm, arid regions.

2. Chardonnay

Chardonnay is one of the most popular wines and its grapes are grown worldwide. Chardonnay can grow in a variety of climates and is grown in California, Italy, Australia, New Zealand, parts of South America, and other regions. Chardonnay grapes are green in color, and yield a white wine. The Chardonnay grape is quite neutral and is mostly influenced by terroir and oak. Chardonnay is commonly used to produce wines of different styles. Ice wines, Champagne, and Pinot Noir commonly will contain Chardonnay additives. Chardonnay aged in oak will be influenced by vanilla, smoke, clove, and cinnamon notes. In cooler climates, Chardonnay produced is mostly light to medium bodied, and more acidic. Flavors include apple, pear, and green plum. In warmer regions, Chardonnay produced has more tropical flavors. Citrus flavors, fig, banana, mango, peach, and melon are noticed. Chardonnay wines that go through malolactic fermentation will have less acidity, buttery, and hazelnut flavors.

Chardonnay is a wine preferred by winemakers for it's adaptability. Often indiscrepancies in Chardonnay can be masked by being aged in oak. Chardonnay vines also yield higher amounts than other varieties, even making a bad harvest cooperable. In hot climates, winemakers may struggle to keep natural acidity. When acidity is lost in Chardonnay grapes flat, flavorless wines result. Winemakers can correct this by a simple addition of tartaric acid (acidification), by harvesting early, implementing oak, and malolactic fermentation. In cooler climates vintners struggle with early blooms which make spring frost a large threat. This can be mitigated by placing braziers between the vine rows.

3. Pinot Gris

Pinot Gris is a white wine commonly referred to as Pinot Grigio (it's Italian name), Rulander, or Grauburgunder. In the United States, Pinot Gris is primarily produced in Oregon, Washington, and California. Pinot Gris is similar to Pinot Noir, and is indistinguishable in a vineyard until veraison (mature ripeness). Orange, pale pink, grey, and dusty purple are the colors used to describe the grape. Common flavors are pears, apples, stone fruit, tropical fruit, and hints of smoke but will depend on the region in which the wine was produced. Dry and sweet Pinot Gris are produced. During fermentation, little no no flavor manipulation takes place. Often vintners will use neutral barrels to subdue the addition of any notes during fermentation. In addition, lighter styles will go through malolactic fermentation. The finest Pinot Gris is mostly derived from cooler climates. Cooler climates support the grape's low acidic, high sugar properties. Although Pinot Gris is sometimes produced in warm regions, vintners are up against low acidity, low structure, and higher alcoholicity sensations.

4. Syrah/Shiraz

Syrah and Shiraz are referred to interchangeably. Shiraz is the Australian name for Syrah, and with Australia being a large producer, both Syrah and Shiraz can be used to refer to the same variety. Syrah was born from crossing the two varieties: Dureza and Mondeuse Blanche. The red and white wine blended help soften the tannic and create a more balanced flavor. France, Australia, and Spain are well known producers of this blend, with Argentina, South Africa, and the United States growing in popularity. In the United States, California and Washington state are the main producers. The grapes appear black rather than red on the wine. Blackberry and plum are common notes associated with the wine. Syrah takes on notes of black pepper, rosemary, tar, smoke, and violets. As Syrah ages, unique sensations such as leather, tobacco, and smoked meats emerge. Tannins are typically high in this wine and acid typically ranges from medium to low. Oak is commonly used to age Syrah and is responsible for influencing the development of mature flavors. Vintners often cold soak Syrah grapes for days and up to weeks (prolonged maceration) which increases color and fruitiness, while limiting tannin and harsh herbaceous notes.

5. Zinfandel

Zinfandel has made a name for itself in the United States, particularly in California. Napa and Sonoma counties in California are known for producing world-recognized and winning Zinfandel wines. Zinfandel is arguably an American native although, some argue, with origins in Italy. Zinfandel is sometimes referred to Primitivo, the Italian version. Either way Zinfandel has raised to become a signature wine of the United States and has grown immensely in popularity. Along with the United States, South Africa and Australia are also producers, although Zinfandel is not a key player in these regions.

The Zinfandel grape appears deep purple, almost black in color. However, the grape is primarily used to create a semi-sweet blush style, Rosé called White Zinfandel. The grape is able to hold high amounts of sugar which can sometimes translate into alcohol levels of higher than fifteen percent. The final, overall Zinfandel taste will depend on when the grape was harvested and how ripe the grape was. The best climate for Zinfandel grapes is warm, but not too hot as the grapes can shrivel up. Zinfandel is susceptible to bunch rot due to the rigorousness of the vine and their tight-growing, bunch tendencies. Zinfandel is often pricey as the harvest can be tedious for a vintner. Zinfandel grapes ripen unevenly which leads some vintners to hand harvest, examining every bunch over the course of weeks. Other vintners opt for harvesting the whole bunch to save time and money.

Vintners are able to manipulate the final product through fermentation length, maceration length, oak aging, and time of harvest in terms of Brix. The brix for Zinfandel typically ranges depending on the type and style but falls between twenty and twenty-five. Tobacco, apple skin, strawberry, and blackberry are the notes that develop as the Brix heightens.

6. Pinot Noir

Pinot Noir is well known in the United States, but produced worldwide. Oregon, California, Washington, Michigan, New York, France, Germany, South Africa, Australia, and New Zealand are top producers of Pinot Noir. Not only is Pinot Noir great on it's own, but is commonly used to add a new element to other wines such as Champagne, sparkling white wines, and others. Chardonnay was actually born from Pinot Noir, and these two grapes are almost always grown next to each other in a vineyard. Both varieties require the same terrier.

Pinot Noir errs on the lighter side of most reds and has an ABV ranging from 11.5% to 13.5%. Some vintners have trouble cultivating suitable tannin levels with this grape. To combat this problem, the vines are pruned immensely to lower the yield. This ultimately produces more concentrated grapes. In addition, many vintners will use whole cluster fermentation to increase tannin and complexity. Whole cluster fermentation includes the entire grape bunch, stems, and some leaves. This technique is rarely used on any other wines and is unique to Pinot Noir. Pinot Noir vines do well in intermediate climates. Long and cool growing seasons help the grape reach full potential. Pinot Noir is often found in valleys or near large bodies or water for an insulating effect. In California, Pinot Noir is often grown in places that receive morning fog and cool breezes to combat the warm climate in California.

Pinot Noir is one of the more difficult grapes to cultivate. The grapes grow in jam packed clusters that require extensive canopy management to avoid disease. In addition, the skins on this variety can be unpredictable, and thin which complicate tannin management. Younger varieties of Pinot Noir boast cherry, raspberry, and strawberry notes. As the wine ages, more complex notes are introduced to add to the complexity. Pinot Noir is often aged in French oak barrels to add more complexity to the wine.

7. Merlot

Merlot is one of the more common wines that can easily be found in a store. In the United States, California and Washington lead the way growing the most. Long Island, New York, Ohio, Texas, Virginia, and Oregon are other states growing Merlot in some capacity. Merlot grapes are some of the most popular in the world. France, Merlot's place of origin, is home to two thirds of the world's total plantings of Merlot (Robinson, 2003). Not only does Merlot have it's own reputation, but is commonly added to other wines to round out the flavor. Many regions growing Cabernet Sauvignon will also grow Merlot grapes, but Merlot grapes typically reside in the cooler portions. Warm regions typically do not harvest Merlot grapes due to their ability to ripen quickly. As it is, Merlot grapes are naturally ready to be harvested much earlier than other varieties.

Merlot can be categorized into three categories: soft, fruit, or brawny. Dark berries, vegetable, earthy, floral, and herbal are some of the words used to describe the different notes of Merlot. Merlot is not always an easy grape to produce and thrives mostly under specific conditions unlike other versatile veraties. Merlot requires cold soils, many of which are clay or clay based. Vintners are up against a variety of concerns when cultivating Merlot. Merlot grapes bud early, and can be susceptible to frost. Because Merlot has a thinner skin, some vintners have to look out for diseases like Botrytis bunch rot and others. Merlot requires well drained soil and extensive pruning. Many believe pruning to be one of the biggest factors in the quality of Merlot. Most vintners will reduce the yield immensely only leaving a few buds to produce higher quality grapes. The age of the vine also impacts Merlot making it hard for newer vineyards to produce high quality Merlot. Merlot must be harvested at the right time leaving little to no leeway. Merlot grapes have the ability to over ripen in just a few days and ruin the whole crop. Some vintners will combat this by harvesting early which also helps the grape retain acidity. Others will harvest later after the grape is over ripe and play off those flavors. Either way, it can be a tricky business.

8. Riesling

Riesling is known as a German wine, but it is also grown in other places of the world and the United States. The Finger Lakes region of New York and the Pacific Northwest are places in the US where Riesling is grown. Riesling is a white wine, considered alongside Chardonnay and Sauvignon in terms of quality. Quality Rieslings are highly sought after. Dry, semi-sweet, sweet, and sparkling wines can all be produced by Riesling. Riesling is highly expressive and showcases the region it was grown in when tasted. The place of origin will highly affect the flavor of the Riesling. Cooler climates produce apple, tree fruit notes, and higher levels of acidity. Warmer climates produce more citrus and peach notes as the grapes ripen. It's mainly thought that it's best to grow Riesling in climates that allow long and slow ripening. Proper pruning is necessary to lower the yield and keep the berries concentrated. The acidity found in Riesling allows the wine to age well over time. Many well-made vintages will express notes of smoke and honey.

During harvest, Riesling grapes require more attention than other varieties. The skin of the grape must be treated gently to ensure that no bruising or crushing occurs. When Riesling skin breaks prematurely, tannin production begins and throws off the balance required for Riesling. Riesling grapes are chilled often during the vinification process unlike other grapes. After the grapes are picked, they are usually chilled for the first time to preserve the delicate flavors that come just after harvest. The grapes are often chilled again after being pressed and again before fermentation.

A bladder press is most commonly used due to their gentleness. While fermentation takes place, the grapes are often cooled in stainless steel tanks. The tanks typically are kept cooler from fifty to six-four degrees Fahrenheit. For contrast, red wines are typically kept in temperatures ranging from seventy-five to eighty-four degrees Fahrenheit. Unlike many other white wines, Riesling does not participate in malolactic fermentation. Riesling is praised for its tart, acidic flavors.

9. Sauvignon Blanc

Sauvignon Blanc is grown worldwide. California, Washington, and New York state are the two most notable states for Sauvignon Blanc production. Ohio is an up and coming producer of Sauvignon Blanc. Climate is the main determinant of the final taste of Sauvignon Blanc. Sauvignon Blanc can take on aggressive, grassy flavors or be sweet and tropical. Cooler climates producing Sauvignon Blanc will yield berries higher in acidity. Notes of grass, green peppers, passion fruit, and floral are most common.

Warmer climates produce more tropical notes but struggle with over ripening. Grapefruit and peach flavors are commonly noticed in Sauvignon Blanc produced in warm regions. Sauvignon Blanc is known as a young wine and is not typically aged. If Sauvignon Blanc is aged, it's typically done so in oak.

One of the major decisions a vintner must decide when producing Sauvignon Blanc is the contact the grapes have with must. Prolonged exposure to must sharpens the intensity and pungency of the wine. Other vintners leave smaller amounts of must for blending purposes. Some vintners will ensure there is no must contact to reduce the aging ability of the wine. The temperature of fermentation is also an important decision a vintner must make. Warmer fermentations ranging from sixty to sixty-five degree Farenheit bring out the mineral flavors in Sauvignon Blanc. Cooler temperatures promote more fruity and tropical notes. Many vintners will use stainless steel for fermentation to keep flavor intensity and a sharpness to the final flavors. Other vintners seldom use malolactic fermentation or oak aging.

10. Rosé

Rosé wine has exponentially grown in popularity over the years, especially with the younger generations. Rosé is one of those wines that has been "trending" for some time. Many believe that a Rosé is the result of mixing red and white wine, although this is highly frowned upon by the winemaking community. In some places, mixing red and white wine to produce Rosé is illegal! Rosé gets its characteristic pink color from limited juice, skin contact. Because wine's receive their color from skin contact, Rosé is one of those wines sitting in the middle of white and red. Rosé is made by juicing red grapes and only allowing a short period of contact. Two to three days is usually the time period. After the short period of soaking, the skins are removed and the juice continues to ferment. In Europe Provence, France is known for their Rosé. In the United States Rosé is produced in New York, New Mexico, Oregon, Idaho, and California. Long Island and Sonoma are some of the biggest names in terms of Rosé in the United States.

Because Rosé is produced by many different red grapes, tons of variations are a result. Rosé is mostly produced by Pinot Noir, Syrah, Merlot, Malbec, Cabernet Sauvignon, Temperanillo, and Zinfandel. Rosé wines can range from sweet to dry, and sparkling to still. Most Rosé is categorized as light and fruity.

11. Malbec

Malbec is a variety of wine grown in almost every region of the United States but it's origins are said to be from Argentina. California, Oregon, Texas, Michigan, New York, New Jersey, North Carolina, Colorado, Missourim and Georgia all have plantings of the variety. Malbec is known for being quite dark in color, taking on inky characteristics. Malbec is typically dark and robust, containing more tannins than most other wines. Malbec is often blended with other varieties to create a more powerful variety. Malbec has thick skin and typically requires more sun and heat than other varieties, such as Merlot or Cabernet Sauvignon. Malbec is not always trellised, especially in it's traditional locations. MAlbec will be cultivated as a bush cine and kept in low yield for more pungent berries. Malbec produced in this fashion tends to be rich, dark, and juicy. The flavors and aromas of Malbec can be described as hints of tobacco, garlic, raisin, and plums.

Vintners growing Malbec have to pay attention to diseases, frost, rot, and mildew more commonly than other varieties. Some vintners will plant a similar variety, or clone, to combat this problem. Extremely warm climates have to pay attention to the acidity of the wine as it can get too low. Malbec is able to grow in many different soil types but thrives most in limestone based soils. High altitude is also a factor in some of the best quality Malbec. The yield is typically kept low for Malbec. Higher yields can produce wines that are flabby and weak.

12. Moscato

Moscato is a grape used to make many different types of wine that range in color. White Muscat Ottonel, yello Moscato Giallo, and pink Moscato rosa del Trentino are a few examples of the different varieties produced. Moscato in particular is made from Muscat

Blanc grapes, although the Muscat variety produces other types of wine. No matter the color, Moscato typically retains nearly the same flavor profile. Muscat grapes are known for their sweet, floral aroma. Muscat is a grape that's been around for a long time, and has sparked recent interest in the American community.

Chapter 4: Wine Production from Start to Finish

The following chapter will highlight each process in winemaking. This chapter is meant to guide you through the main processes that vintners go through to create wine. This chapter should be used only as a guideline; keep in mind there are over a hundred factors that can dictate the final quality and taste of wine for any varietal. In addition, this chapter will be generalized, as every winemaker has their own style and way of doing things. Through the chapter you may find processes that require professional help, and that's okay. World renowned wineries often have different professionals in and out of the winery to perfect the process. This guide is meant to help you get started and give you the basic knowledge needed. The chapter will begin with what's required before planting and continue all the way until being bottled. Winemaking is not a fast process which you will see after reading this chapter. Many steps and small requirements are necessary to end up with a wine worth drinking let alone selling, so be patient!

The Basic Processes, In Steps

The following section will go through every process a wine maker should go through from start to finish. The guide is meant to inform you of the basic practices and options available. There are many different choices to make when growing and processing grapes. Remember to use your own discretion as each variety and climate requires the practices to be tweaked.

1. Test the Soil

The first step is to evaluate the soil you're going to be working with. To do this, take a shovel and dig a large hole up to three feet to obtain an accurate sample of subsoil. Scrap the soil off of one side of the whole and add to a ziploc bag. Then, scrape the soil from the bottom of the hole (1 to twelve inches) and place it into a ziploc bag. Third, scrape soil from two feet deeper and place it into a ziploc bag. Be sure to accurately label each bag.

The next step is contact your local Agriculture Extension Office. This is where the soil can be tested in a laboratory type setting. Soil samples will alert you of nutrient problems before planting, which is key. Seven is the optimal pH, which is neutral. The laboratory can also alert you of soil infested with nematodes or root-louse phylloxera. If the soil contains any of these pests, you'll need to choose a rootstock that is resistant to these pests. When ordering from a nursery, you'll need to specify your desired rootstock and vinifera. A nursery can graft and combine almost any combination imaginable. If the soil sample is free of these pests, it's possible to plant the vines using their own roots. More research may need to be conducted for threats in your area.

2. Irrigation

The next factor to take into consideration is the water situation. Research your climate and document how much rainfall there is in a year, and when. Applying a water system through the ground or by drip irrigation will be more efficient than installing sprinklers. Water should be kept off the fruit and vines to avoid disease. Remember that wine grapes should only receive just enough water to survive.

3. Crop

If you're just starting a vineyard, you'll need to order crops. It's estimated that five pounds of crop per plant requires 200-250 cines. This example ensures enough fruit for one 60-gallon barrel of wine and extra for topping. The ratio of 250 vines to 60 gallons is a good place to start when figuring out yield. Keep in mind that different soils and crop levels will produce different yields. Also keep in mind that 20 pounds of fresh fruit is about equal to one gallon of homemade wine (Hagen, n.d.). For example, if each vine produces five pounds, and desired yield is one five-gallon batch, at least twenty vines should be planted.

4. Trellis

The type of trellis and style should be decided upon before planting grapes. The best way to test what works best without an expert opinion is to order a few vines a year and test the plot. Vines and grapes will react differently to different soils, climates, and water. Low vigor sites typically implement a VPS while high vigor sites may need sprawl on a common wire trellis. Vines can be trained but your trellis system should complement the variety of grape and natural tendencies.

5. Planting, Irrigation, & Planning

Before winter there are a few things to do to prepare your soil. The first instruction is to break up the soil. Break up the soil and go as deep as possible. Loose soil promotes deeper vine growth from the vines which is desirable. The deeper the root system, the better nutrients the plant receives. Breaking up the soil will get you accustomed to any problems beneath the surface. Extensive clay or restricted soil should be dealt with by placing small stones for better drainage. Although this is not a necessary step, it may be helpful for drainage. Many vineyards will stack small, light-colored rocks at the base of the vines. This works as an insulator on cool nights by reflecting light into the canopy and soil for warmth. This is not necessary but will help the fruit mature, which is helpful for those in tricky climates.

The next step is to plant a cover crop. There is a specific type of cover crop plant for every soil type so be sure to purchase the right one. Clovers, subclocers, vetch, and other plants are examples of cover crops that aid in regulating the nitrogen levels in soil. Rye, barley, and other grasses are sometimes used as well for erosion purposes during seasons with high rainfall. Some vintners chose to plant different types of flowers to control insects on vines. The idea is that the pests will be drawn to the flowers, rather than the vines. The perimeter of the vineyard can be planted with marigolds for pest control.

During this time, survey the land for animals. Gopher, voles, moles, deer, pigs, rabbits, and other animals may have established homes on the property. Some vintners will combat animal presence by installing fences around the property. Some will keep dogs and cats living on the property to scare animals away; so, don't be afraid to bring your animal to work. Bird netting is another option some take advantage of, since birds will be attracted to the fruit as the grapes sweeten.

When ordering crops, it's important to have the best variety selection and rootstock selection for your area and climate. Order the vines ahead of planting to be sure you have everything you need. Delayed planting may create more work in the future. Amend the soil as necessary. A local agricultural professional can help you choose any soil additives to balance the pH and substances in the soil.

The soil should be balanced as follows: Nitrogen: 4 to 8 ppm, Phosphorous: 30 to 75 ppm, and Potassium: (exchangeable) 81 to 500 ppm (Hagen, n.d.). Amendments should be mixed into the top two feet of soil. The water will later carry them down further to the roots. Vintners will sometimes use a fungus called mycorrhizae to help young vines get the nutrients they need. Chemical solutions are not suggested as they can throw off the natural microbial balance of the soil.

In regards to irrigation, a system should be planned out to keep as much water off the leaves and grapes as possible, as mentioned previously. Small plantings will benefit from a furrow dug around the vine row. The furrow should be flooded every few weeks using a hose. For a more complex system, a drip system can be implemented. The first step to a drip system is attaching a hose bib to black irrigation tubing. The tubing should run the length of the vine row on the lower trellis wire. One end of the tubing is closed shut. Drip emitters are then attached along the hose. A pressure compensating emitter will ensure each is dripping equal amounts onto the slope. It's suggested to use one gallon per hour emitters on flat vineyards and thirty percent more for vineyards with higher slopes (Hagen, n.d.). When the vines are growing, five gallons a week is all the vine needs to be healthy. During rainfall turn off the irrigation system.

6. Vine Training

Vine training refers to the support system of the plant. The training system chosen will depend on historical tendencies, legalities, climate, and practicality. Training systems are defined in many different ways such as by the height of the vines, by the type of pruning, and by the type of trellis used. There are a few different types of vine training.

Bush-trained vines are usually standing on their own, or supported by one or more stakes. In the past, bush training was the most common practice used. As technology progressed, bush training became less suitable because it does not work well with mechanized machinery. A compact, sturdy plant in hot and dry climates can benefit most from this type of training. For this system, vine arms are spur pruned. Extra ripeness can be achieved using this system if the bush is kept close to the ground.

Replacement cane training uses a wire trellis system for support. One or more canes will run along the wires depending on the system used. The most popular system is called the guyot system where one or two canes are tied to the bottom of the trellis.

Cordon training has many different variations but the sample principle. The trunk of the wine will be extended horizontally across the wires. The cordon is close or high from the ground depending on the system used. Cordon systems work well with operated machinery that aids in pruning and harvest. This technique is also helpful in vineyards that have strong winds.

Vertical Shoot Positioning or VPS requires four pairs of wires on a trellis. The vines are trained vertically when this system is used. The VPS system has many variations and may implement the ideals of the other training systems mentioned. In this system, often the cordons are trained in opposite directions along the bottom wire. This system allows ten to twelve spurs on each cordon but they are later pruned. Because the vines are not as close to the ground, growing vertically, the chance of disease is lessened when using this system.

7. Maintenance & Pruning Methods

In every vineyard, pruning and canopy management is needed in order for the grapes to be of best quality. Pruning helps control the shape, canopy, yield, and vigor of the grape. Important pruning takes place in mid to late winter. Pruning cuts away much of the growth from the previous year and gets the vines ready for a new yield. For a replacement cane system, two canes will be left with only six to twelve buds depending on the variety grown and the winemaker's preference. For the bush and cordon system, short spurs will be left and with only about two buds. Canes and spurs refer to the previous year's wood. Grapes only grow from the flowers of the previous year's wood. The rest is pruned because it promotes better quality grapes.

It is generally agreed upon that lower yields produce better quality grapes but some argue canopy management is more important. When pruning, start at the trunk and move outwards eliminating dead canes. Then remove canes that branch from another cane. The canes kept should originate from the cordon. Now, count the canes remaining. The number of canes left should be decided upon in accordance with the variety, maturity of the canes, and vintner's preference. The canes should be healthy and evenly spaced. Trim each cane three to five nodes depending on the variety being grown. A half inch should be left below the last node. The spurs should also be evenly spaced and pruned to one node extending from the cordon. Lastly cut extra canes flush with the cordon.

8. Harvest

Knowing when to pick or harvest the grapes, is one of your most crucial decisions. Harvesting too early or too late will affect the final outcome greatly and could even ruin the entire wine. An experienced vintner will be able to taste a grape and receive valuable insight. However, other techniques are more commonly relied upon.

The first thing to look at is the amount of sugar and acid in the grapes. This can be done by picking different grapes for one variety throughout the vineyard, and squeezing them into a sample. This practice is useful for large vineyards to get an average reading. Smaller vineyards with less land may choose to only sample a few grapes due to similar conditions throughout the limited space. Although each grape may have a different Brix level the overall average is most important. Typically, the brix will range from 21 to 25 at harvest depending on the variety. The pH level for white wines at harvest is typically from 3.1 to 3.3. Red grapes will have a pH range of 3.3 to 3.5 at harvest.

Titratable acid is an acidity measurement and typically ranges from 0.6 to 0.9 per milliliter. Titratable acid expresses all of the acid in wine with the most dominant being tartaric acid. As a grape ripens, the pH will drop, naturally. The level of pH can be determined by using a pHmeter. As you know, acid is important in the winemaking process and final product in terms of color, balance, and taste. A helpful table is given below to familiarize yourself with the suggested levels according to Rick Haibach (2017), an award making vintner.

Red	Harvest pH	Brix (in degrees)	Titratable Acid (per milliliters)	Final pH
Cabernet Sauvignon	3.3-3.4	24-26.5	0.6-0.7	3.6-3.7
Merlot	3.2-3.4	23-25.5	0.65-0.8	3.55-3.6
Malbec	3.2-3.4	23-26	0.65-0.8	3.55-3.65

Cabernet Franc	3.2-3.4	23-25.5	0.65-0.8	3.55-3.6
Pinot Noir	3.2-3.3	22-25	0.65-0.8	3.5-3.55
Zinfandel	3.3-3.34	24-28	0.6-0.75	3.65-3.75
Average Red	3.2-3.4	22-27	0.6-0.7	3.55-3.7

White	Harvest pH	Brix (in degrees)	Titratable Acid (per milliliters)	Final pH
Riesling	2.9-3.2	20-24	0.7-0.9	3.1-3.4
Gewurztraminer	2.9-3.2	20-24	0.7-0.9	3.2-3.4
Sauvignon Blanc	2.9-3.3	20-24	0.8-0.9	3.2-3.4
Pinot Grigio	2.9-3.2	20-24	0.7-0.9	3.2-3.4
Chardonnay	3.0-3.3	22-25	0.7-0.9	3.3-3.45
Rosé	2.9-3.3	18-23	0.7-1.0	3.2-3.5

The second decision that goes into harvest is impending weather. Unlike other fruits, grapes do not continue to ripen once picked. If the grape is picked too early, the grape may be too tart or sour. If the grape is picked too late, the grape may not contain the necessary acid and be flabby. It can be difficult during abnormal weather seasons to assess when to pick the grape, so keep those factors in mind. Overall weather patterns are similar year to year in one location, but monitor unseasonable weather predictions around harvest time. Unseasonable high or low temperatures, rain, hail, or frosts are threats to the grape. Temperature may be a determinant to harvest earlier or later than the year prior. Hot seasons can quicken grape ripening while colder temperatures may prevent the grape from maturing. Each season may require different growing periods due to the weather.

The majority of harvesting in the Northern Hemisphere will fall between the months of late August to early October with a few outliers. Each variety will vary but grapes typically need 150 to 180 (frost-free) days of veraison from spring to fall (Moon, n.d.).

When all of the factors are aligned, a vintner will decide to pick the grapes. Usually the vintner will start with grapes most exposed to the sun or in a higher nutrient soil area in the vineyard. Grapes in these conditions ripen quicker than others in shaded, cool, low nutrient soils. Individual rows may be picked instead of all at once. Most wineries producing high quality wines prefer handpicking the grapes. This allows the worker to pick grapes that are ready, or leave and discard undesirable looking grapes. Unripe or diseased grapes are left behind which produced a better, final product. Workers will use sharp shears and a bin to collect the grapes. When the bin is full, the grapes are transported onto a tractor and delivered to the winery. Vineyards that specialize in mass production may choose to use machinery when possible to pick the grapes to cut down on time and costs. With this method, more time is needed to sort through the grapes once they arrive at the factory. This increases the chance of missing a bad bunch since the grapes are only looked over once.

Technology has improved, but machinery leaves more room for error and sorting becomes a strong focus. Using machinery also makes the grapes more susceptible to getting broken or pierced as grapes are removed by shaking, batting, or stripping the vine. Most vineyards that use machinery must harvest at night to lower the chances of the grapes fermenting too early due to broken or pierced skin. Harvest can last from one week to over a month depending on the methods used and size of the vineyard.

9. Processing, Crushing, & Pressing

When the grapes arrive at the winery, the next process is called *triage* which comes from the French language. This process starts when the grapes are put onto a sorting table. Workers look over each cluster for quality purposes and remove unwanted grapes. Unripe, diseases, and damaged grapes are discarded along with any leaves or bugs. The grapes are not washed, and never are throughout the entire winemaking process. Although some wineries sort by hand, some take advantage of sorting technology. However, most vintners prefer sorting by hand unless they can afford top quality machines, to eliminate error.

After the grapes have been sorted, the destemming and crushing occurs, likely at the same time using one machine. Most destemmers can remove stems before or after the grapes are crushed, based on the vintner's preference. Tannins are the main reason for destemming the grapes. Modern machines usually use a large steel or aluminum trough with a screw, likely rubber, on the bottom. As the screw turns the grapes are gently pulled from the stems. The grapes must be gently destemmed to avoid damaging the seeds in the grape.

In some situations, some vintners may include a particular amount of stems in the fermentation process by leaving some on or adding them back in. Stems, skins, and seeds contribute to more structure, texture, and bitterness in wine. However, keeping too many stems in the must during fermentation may yield a wine likely too tannic for consumption. Because grape skins contain suitable amounts of tannins, many winemakers will not opt to keep any stems at all but this all depends on the wine being produced and the vintner.

The grapes are crushed into a liquid called *must* to trigger fermentation. Crushing is different from pressing. Crushing merely breaks the skin open while pressing removes as much juice as possible, leaving behind any solids. Crushing is necessary to get the juices flowing and begin fermenting. The sugar and yeast begin to interact. The yeast is responsible for turning the sugar into alcohol and carbon dioxide. The seeds play an important role.

10. Pressing

Pressing is the process of juicing the grapes. Skins, seeds, and other solid are left behind. Pressing occurs at different times depending on the wine being produced. For white wine production, pressing typically occurs directly after crushing, before fermentation. When producing red wine, pressing typically happens after or near the end of primary fermentation. There are many different types of presses that can be used based on a variety of factors.

Although, a wine press can be categorized into two categories: continuous press or batch press.

A continuous press may come in different variations, but most rely on the same concept. Grapes are placed into one end of the press and carried along the length, and undergo various pressures along the way. The vintner has the option to collect must at many different points while the grapes undergo the pressure. Typically the best juice comes from the early collecting points. However, this technique is mainly used in large industrial settings and the quality of the grapes is inferior to those that undergo other methods of press. A continuous press is helpful for processing high quantities of fruit, but is not the preferred method for most craft vintners.

The second category of press is batch press and can be further categorized into three types: horizontal plate press, horizontal pneumatic press, and vertical basket press. The first, horizontal plate press, is sometimes referred to as the Vaslin press. The horizontal press works by using a perforated cylinder. Inside of the cylinder contains plates, hoops, and chains. Grapes are placed into the press through a hatch, and the press rotates. This tumbles the grapes releasing free-run juice. Next, the metal plates move the grapes to the center of the press and the plates squeeze them further. The juice released by the metal plated is typically more concentrated. Afterwards, the same process is repeated until the grapes contain no more juice. Each time this method is repeated on the grapes, a lower quality juice results. The best juice is produced after the grapes go through the process once. The same batch of grapes may undergo this method two to four times. A horizontal press is usually affordable and used in smaller winery settings. Smaller properties making red wine generally favor this type of press.

The second type of batch press is the horizontal pneumatic press which is sometimes referred to as a Willmes or Bucher press. Pneumatic presses are favored for white wine making. The pneumatic press uses a gentle process and uses low pressure. The press operates by rotating and using an inflated central bag, bladder, or sheet. Compressed air flows in and out of the press through the bladder which presses the juice out of the grapes. When the central bag inflates, the grapes are pressed up against the side of the press and the juice flows to a controlled cylinder. This process happens numerous times with the same batch of grapes, but the best quality juice will result from the first press. In some scenarios, a pneumatic press has the ability to eliminate oxidation risks by flushing the tank with nitrogen or carbon dioxide. Many pneumatic presses in modern times are controlled by a computer and can cost over $50,000 USD.

The third type of batch press is the vertical basket press. The vertical basket press has been around longer than most other types of presses, and is referred to as the original, traditional press. Many of the world's finest wineries still use this press today, using original methods. The vertical basket press is known for being gentle and is a very slow process. Red and white wines can be pressed using the vertical basket. The basket can handle whole clusters or must, but the solids must be manually removed from the press after each operation. In some situations, the vertical basket press will allow the wine to be static, and less bitter then when using other press methods. Spicy wines benefit from this type of press because the notes are supported. The vertical basket press operates by lowering a wooden plate on the grapes sitting in the basket. The basket can be deep or shallow depending on the vintners preference. The basket is typically made of wooden staves, although nowadays there are different variations.

11. Fermentation
a. Temperature

Temperature is important throughout the entire fermentation process. During fermentation, chemical reactions are occurring and temperature can fluctuate naturally. Temperature should be monitored and sometimes manipulated for this reason. Temperatures may start low but rise during fermentation with over a twenty degree (farenheit) difference. As temperatures rise, yeast functions differently. If temperatures reach over ninety to ninety-five degrees Fahrenheit, the yeast will stop working. This is sometimes referred to as stuck fermentation, and in most cases is not desirable. This can be detrimental if sugars are not fully fermented.

Temperatures are manipulated throughout fermentation to receive different results, but is a tricky process. Warm fermentations produce great colors and speed up fermentation. Cooler temperatures aid in growing yeast and produce higher alcoholic degrees. For example, a vintner may start fermentation at a cooler temperature and allow the temperature to rise naturally up to a certain point. The vintner may choose later to lower the temperature again to ensure total fermentation and complex compounds. Some underground, cool cellars have the advantage of being self regulating. While other wineries pump wine through heat exchangers to higher or lower temperature. Tanks may be hosed on the exterior with cool water or wrapped in cooling jackets. In other cases, temperature controlling devices may be inserted into the actual tank.

b. Extraction and the Cap

Extraction refers to the process of drawing out flavor, tannin, and color from the grapes throughout the entire winemaking process. Extraction is a tricky business and completely depends on the vintner's style, preference, and the variety of wine being produced. For red wines, fermentation usually takes place in open vats. While in the vats, the solids and skins rise to the top due to the release of Carbon Dioxide. The visible solids seen at the top of the vat is referred to as the cap. There are many different practices for managing the cap. Most vintners want the solids and wine to be well integrated to aid in extraction while fermenting. In addition, the cap can promote bacteria and spoilage due to warm temperatures and an environment suitable for growth. Remontage is used by vintners to combat this problem.

Remontage is the process of drawing juice from the bottom portion of the vat and spraying the juice over the top of the cap to integrate it back into the juice. This is important for extraction, fighting disease, and accelerating total and complete fermentation. Other techniques used involve punching down the cap which is a gentler process. This is called pigeage.

Wooden paddles, sticks, or sometimes feet are used to perform pigeage although some vats have this technology built in. Rotary vinifiers work to continuously tumble the grapes increasing the skin to juice contact. Punching down or pumping over is not needed when this technology is implemented. This technique will quicken fermentation and is not suitable for producing all types of wines. Rotary vinifiers are mainly used when producing soft, easy to drink red wines.

c. Maceration

Maceration is the process in which the phenolic compounds of the grape (tannins, colors/anthocyanins, and flavor) are released from the grapes and transferred into the must. This occurs by soaking the grape solids in the juice. Maceration can take place during different parts of the winemaking process and varies from vintner to vinter or by the variety being produced. Pre-fermentation maceration or post-fermentation maceration are the two decisions vintners need to make. In some scenarios, maceration may be skipped all together by the vintner if maceration won't be beneficial. In some cases, maceration can cause hard tannins to be extracted, so the vat is drained directly after alcoholic fermentation. Cold soaking or pre-fermentation maceration works to promote fresher, cleaner wines. The technique was introduced by Guy Accad, an oenologist from the 1970s. Accad believed maceration before fermentation would yield wines with greater finesse and top wineries have shown this to be true for some styles. Post-maceration can last anywhere from two days to an entire month based on the vintner's decision. Bottles that are destined for long aging periods will benefit most from post maceration. If the wine is not destined for a lengthy aging period, producers often cut post-maceration short.

d. Racking

Racking is the process of riding unnecessary sediment or lees. Lees is often the term used to refer to sediment such as dead yeast cells. Gross lees are coarse and settle at the bottom of a vat after fermentation. Fine lees often settle after an initial racking and may need to be removed during a process called fining. Racking is necessary for achieving clear wine and to reduce the risk of rancid flavors. Racking transfers juice from one vessel to another, which leaves sediment behind. Racking happens a few times during the winemaking process and is up to the discretion of the vintner.

e. Malolactic Fermentation

Malolactic fermentation or MF for short, is sometimes referred to as secondary fermentation. The MF process deals directly with bacteria and has little to do with yeast. Malolactic fermentation begins when primary fermentation is complete, around zero degrees Brix. Lactobacillus, Leuconostoc, and Pediococcus are the genera of bacteria involved in MF. Malic acid can be compared to the harsh, sour taste when eating a granny smith apple. Malic acid is converted to lactic acid during MF. Lactic acid has a much softer taste and can be compared to sensations when drinking milk. If a winemaker does not wish to use MF, sulphur dioxide and colder temperatures are used for prevention. MF is though necessary for many varieties of wine in order to prevent fermentation from continuing once bottles.

Chardonnay is an example of a wine that works well with MF. Riesling and Sauvignon Blanc are examples of wines that are valued for their acidity and are typically not put through MF, as MF would decrease valuable flavors.

f. Maturation

Maturation is the time spent after primary fermentation to finalize the flavors and aromas of the wine. Immediately following fermentation, many wines have not reached optimal quality. Maturation allows the flavors to come together and create a better tasting wine. Tannins soften and acidity levels balance out during maturation. Because of this, different approaches are used by vintners. The time and vessel used to mature a wine is strategic. A vintner will decide both elements based on expertise and knowledge. Time and the vessel can influence a wine for the better, but can also ruin a wine. Maturation vessels include stainless steel vats or wooden barrels. The type used will mostly depend on budget and the outcome desired.

Some vintners chose to skip maturation all together, but this is mainly done with industrial, mass produced wines. Stainless steel is chosen by some vintners for its ability to resist oxygen. Stainless steel is impermeable and can hold wines for a long period of time before bottling. Many inexpensive wines are held in stainless steel until ready to be bottled or ordered by producers. Barrels are typically the preferred method of maturing wine. Most high quality red wines will spend some time in a barrel ranging from nine to twenty-two months. Others will be kept in a barrel for years which the final price will definitely reflect.

While maturing, wine takes on new flavors and aromas as a result of being in contact with the oak. Wood tannins and vanilla are two examples. The size of the barrel also plays a role on how much time the wine needs to mature. Smaller barrels promote faster maturation. Smaller barrels will influence the wine more than a bigger barrel because the surface to wine ratio is smaller. The type of oak used will also influence the wine, with many different variations all ranging in prices from high to lower. Because oak aging is quite expensive, some wines are matured in vats, but oak chips are added for influence. Once the barrels are filled, a mallet is commonly used to tap the barrel and rid air bubbles. Oxygen is not preferred during maturation and vintners will do their best to eliminate any oxygen in the entire maturation process. Many vintners "top" the barrels to make up for any wine loss during racking or due to evaporation. However, topping is sometimes avoided if the container is secure enough.

12. Bottling
    a. Fining

Fining is the process where a substance is added to wine to absorb or rid smaller particles by ultimately creating large particles, making removal easier. Lighter matter, known as colloids, are able to pass through filtering. Colloids are not desirable because they create a hazy appearance and later form a deposit. Because colloids are charged, a compound with the opposite charge can be added to remove the colloids. This is done by using fining agents such as egg whites, gelatine, or bentonite to name a few. This is a tricky procedure as the fining agents themself can hurt the wine more than help if not done correctly.

b. Filtration

Filtration is the process used to rid solid particles in the wine, and takes place throughout different stages of the winemaking process. The three different categories of filtration are: Earth filtration, Sheet filtration, and Membrane Filtration.

Earth filtration is the technique commonly used first to remove gummy solids resulting from dead yeast cells and other solids from the grapes. This method takes place in two stages. First, kieselguhr or earth os placed onto a screen within a filter tank. Then, more earth is mixed into the wine to form a slurry. The slurry is continuously placed on the filtration surface to replenish the wine as it passes. Eventually the system will clog and the process is repeated. The system used contains a vacuum like filter that removes the sediment.

The second type of filtration is called sheet filtration; also referred to sometimes as plate and frame filtration. Perforated steel plates are held in a frame. Then, sheers of cloth or paper are suspended in between the plates. The plates squeeze together mechanically or by using hydraulic methods. Wine is then pumped through and passes through the filter. Matter is trapped into the paper or cloth in the filter and sediment is removed.

Membrane filtration is the last method and is mostly used before bottling if being used at all. By this time, the wine is already quite qualified through the other filtration methods. Wine is filtered through this system under immense pressure which ultimately separates even the finest particles from the wine. Because this process can be intense, not all wines are suitable for membrane filtration. Full-bodied red wines, for example, likely will not use this method as it can trip the body and flavor.

c. Stabilization

Stabilization methods are used to prevent the wine from forming tartaric acid once bottled. Tartrate crystals can form after long periods of storage, or if the consumer places the wine in the refrigerator. These crystals can look like glass and cause panic in a customer. Tartrate crystals are derived from potassium or calcium salts. Although harmless, methods can be used to avoid access tartrate crystals from forming. To inhibit the crystals from forming, the wine is cooled to 25 degrees Fahrenheit or lower and kept chilled for over a week. After, the crystals will appear and the vintner is able to remove them before selling.

In addition, other treatments and starlization are required prior to bottling. The most common treatments available to a winemaker are cold sterile filtration, thermotic bottling, flash pasteurisation, and tunnel pasteurisation. These methods are to ensure the wine is stable and ready to go.

The first, cold sterile filtration, is a method where wine is filtered through sheets or membrane filters to eliminate yeast cells. This process does not use heat, unlike the others. Following this process, the wine is bottled in a clean environment. This process is mainly suitable for wines containing residual sugar and medium alcohol. Any other style could risk re-fermentation after being bottled and isn't suitable.

Thermotic bottling is the method of heating wine to a high temperature and bottling the wine while it's still hot. Often wines will be heated to around 130 degrees Fahrenheit and bottled immediately. The high temperature kills any active yeast and bacteria is not able to survive in such high temperatures. Any remaining bacteria is killed by the temperature and alcohol.

Flash pasteurisation is similar but is only heated for one or two minutes. Typically the wine is heated to about 203 Fahrenheit then rapidly cooled. The wine is then bottled cool immediately.

The last popular method is tunnel pasteurisation. In this technique, the wine is bottled cold. The wine is then passed through a heat tunnel. The sealed bottled are sprayed with hot water which raises the wine temperature to about 180 degrees Farenehit for fifteen to twenty minutes.

d. Sulfur Dioxide Levels

Sulphur dioxide is typically used during the winemaking process but also before bottling. Sulphur dioxide is the preferred method used to antioxidant and disinfect wine. However, not all sulfur dioxide acts in this way. The portion that is bound with the wine becomes inactive. Before bottling, winemakers must adjust the effective sulfur dioxide to inhibit further fermentation and spoilage. Free sulfur dioxide levels should be adjusted anywhere from 25 to 35 mg/l before bottling.

e. Closures

Closures are very important for preserving wine. Closures eliminate oxygen from getting in the bottle and spoiling the wine. The three most common methods are cork, synthetic, and metal screw. Corks are the most traditional way of closing bottles. Many winemakers prefer cork for it's natural qualities. For a cork to be effective it must have elasticity. This means that when the cork is pressed, it will expand and create pressure against oxygen flow. Corks are effective but become less effective when dried out. This is why many wines with cork closures are placed on their side to keep the cork moist. The second method of closure is synthetic. Plastic or cork alternatives fall into this category. Plastic stoppers have recently become popular, but are not preferred for top quality wines. Many consumers complain that plastic closures also ruin corkscrews. Synthetic corks made of polymer often have similar qualities to cork. Lastly, the metal screw cap is preferred by some vintners for its ability to block out oxygen. White wines often use metal screw caps for their effectiveness.

Chapter 5: How to Sell Wine

Now that you have produced the most perfect wine, it's time to sell. If you're not able to sell your wine, you won't see much revenue. Don't let this scare you because with the right knowledge and practices, you'll learn how to gain customers and keep them coming back. This chapter will go over marketing practices, pricing, and digital marketing.

How to Market and Increase Sales

Now that you have the final product ready to go, it's time to market your wine. In a perfect world, every person that tries your incredible wine will be life-long customers and buy tons of bottles every month. In reality, many customers may be one-time purchasers. Don't let this discourage you. Sales are sales. The following section will give examples on how to promote your wine keeping one-time buyers in mind. Anyone that purchases your wine should become a lead with the intention of turning them into long-time customers. Exposing your wine using the following methods will get bottles sold asap, with the potential of customers coming back for more.

1. Open a Tasting Room

A tasting room is an excellent way to draw customers to your site and build brand loyalty. Tasting rooms allow customers to sample wines before purchasing. Allowing samples increases the chances of satisfaction from the customer. Time and time again individuals will buy wine from a grocery store, take the wine home, and be dissatisfied. Then the next time the individual goes to the store, they're likely to skip that brand all together and try a new wine, or revert back to a favorite. By allowing individuals to sample before purchasing, you're guaranteeing a purchase of a wine the customer really likes, and increasing the chances of future purchases.

The key to a great tasting room is helpful staff and a great environment. The employees need to be knowledgeable, friendly, and professional to leave a good impression. The staff should be enthusiastic and passionate about wine. In the wine business, it's not always enough to just be friendly, knowledge is key and will set your business apart from others. Hiring and training someone who enjoys wine will not only make their job easier, but will be beneficial to you as many customers ask employees about their favorites. The employees should have a good grip on every wine being sold.

The next step is to decide if you're able to do free tastings, or charge a small fee. This decision should depend on your financial position and the quality of your wine. It's not uncommon for a winery to charge a small fee of up to five dollars, and a higher fee when trying more expensive wines. If you decide to charge a fee for a tasting, do so in a way that is simple. Confusing the customer or having the customer thinking about prices will ultimately distract and take away from the sample itself. Many wineries that charge a fee for tasting will waive the fee if the customer purchases a bottle. Charging for a sample to a buying customer can come off tacky and in bad taste. Decide the minimum a customer must pay to waive sample fees and make sure all staff are on the same page. If your tasting room accepts tips, be sure to list the information so it's visible to customers. A short "tips are always appreciated" near the bar on a board or on a small jar never hurts, but don't make it mandatory.

Printing a list or displaying information is a great idea. Chances are you won't be hiring 50 employees to work in the tasting room at one time. When rooms are busy staff will be busy and may not have time to thoroughly explain answers to questions. Many of the same questions will be asked in the tasting room. Having a brochure or even a chalk board with information near the bar will free up a lot of time for staff. Include information about the winery, it's practices, and common information on the types of wine you sell.

The next step is to create a tasting menu. Tasting menus can either be simple or complex depending on your budget. If your budget is tight, a simple paper will do. Even if you chose to save in this area the menu should be professional looking, list all of the wines, and give a good impression. As your revenue and success takes off, consider designing a fancier tasting menu. The menu should still have all of the same elements as before but can be made of nicer paper, include images, and be of good font. Spending a little more money or time creating a tasting menu may act in your favor because now the customer has a little souvenir to remind them of your experience. If the wine is superb customers will also take notes on the card and save it for reference. Dedication of a few lines under each wine will allow the customer to take notes or write down the different flavors they taste for a more beneficial experience. In addition to listing all of the wines available, be sure to put a small description of each so the customer knows what they're sampling. If your shop will also serve snacks or food, list food pairing on the menu near the wine.

Now that logistics are figured out, brainstorm ideas on how to make your room a great environment to be in. Not only do you want to attract visitors, but having them come back and bring friends or family will boost your business. An indoor-outdoor feel is always appreciated in areas with nice weather, and a good view will have customers spending more time at your spot. As your business grows, turning your location into a hangout spot will grow your repeat customers. Serving good food and wine in an excellent setting will always boost traffic.

2. Wine Club

Many successful wineries create a club that gives members an exclusive deal. Not only does this create excitement, but it has the potential to generate more sales then what would have taken place. How it works is that each quarter customers commit to ten to fifteen bottles of wine but receive significant discounts, and special perks. Wine clubs promote repeat business and can be more appealing to individuals looking for a deal. Because sales and promotions happen in nearly every store and winery, a wine club is a way for you to compete and draw attention. Wine clubs generate a steady flow and revenue you can count on. According to Wines Vines Analytics (2016) a wine club makes up 33% of total revenue for an average winery, which that number increasing for wineries producing less than 2,500 cases yearly.

3. Host Events

Hosting events not only has the potential to generate extra income, but exposes the vineyard. A stunning location, and great experiences will have visitors and guests coming back to explore what your business offers. Weddings, special events, and summer concert series are excellent ideas. In addition, create unique experiences for guests that they can't find anywhere else. A few ideas are blind taste tests, classes from the head vintner, or having visitors participate in the wine making process. Use your business values as leverage, whatever they may be. If you're a family business, consider hosting a family friendly event with activities. If you're an organic winery, teach a class on sustainability in the vineyard. If you're an animal lover, host events that allow animals. All of these events will show who you are as a business and play into the emotion aspect that many are seeking from a wine producer. Each event gives you the opportunity to sell your wine, or generate leads.

4. Pop Ups

The key to this concept is offering samples. Samples are a tool used by many manufacturers and stores that have shown to be successful. Studies show that up to 48% more people will buy a product after sampling

(Pinsker, 2014). Although you'll likely have samples at your winery, think outside of the box! Start thinking of local farmer's markets, festivals, and local liquor stores. Having a booth in each of these settings will absolutely drive up sales. Many will buy a bottle on the spot, or pay the vineyard a visit. These settings are an excellent way of acquiring email addresses in exchange for a free sample. Keep business cards and information on the vineyard handy to give out. Keep in mind that in some of these locations permits may be needed, so check which your local authorities first.

5. Use Influencers

Influences are all the rage in today's day and age, for good reason. Putting aside stigmas, influencers are individuals who spend a lot of time online, building their own personal brand. Along the way, influencers do many paid sponsorships and market products to their followers. An influencer is able to reach a lot of people which can be beneficial to your business. Although influencers usually charge for promotions, someone with a lot of influence promoting your business will generate sales. In addition, consider hiring micro-influencers who have less followers. An individual with around two-thousand followers that are in your target audience could be just as helpful. Micro-influencers usually charge less for promotions and some may even work for free in exchange for product. All in all, influencers can be helpful as more and more people spend much more time online rather than looking at ads in the local paper.

6. Hold Seasonal Sales

Although you may partake in sales all year long, playing up to holidays and seasons will increase your sales. Wine can be very festive and compliment a holiday well. For example, for New Years run a sparkling wine sale. For Valentines Day hold a sale on your signature red and white wine. For Mother's Day and Father's Day, give a special promotion for bringing your parents. For summer holidays run a sale on your most refreshing, chilled wine. Then in the fall play up your deeper, more lush reds. As you can see, wine can easily pair with different seasons and holidays. Holidays are times where many people gather and alcoholic purchases increase.

7. Offer Gift Cards

This idea may seem like a no-brainer but you'd be surprised how much this gets looked over. So often wine is a gift commonly given. Offering gift cards allows the customer to think about who they could give it to. Many struggle giving the perfect gift, and a gift card to a local winery is a great option. Gift cards are highly popular and will draw more attention to your shop.

The Best Digital Marketing Practices

As you may have heard, the digital way is taking over. Many businesses, no matter their area, are converting to online practices to increase traffic. Because the competition in the wine world is stiff, moving your business online or having online features will greatly increase your chances of success. This section will go over the various ways to increase your online traffic.

The first thing to consider for digital advertising is email marketing. Email marketing will appeal to existing and new customers, which makes it a powerful tool. Because most emails are gained willingingly, meaning an individual has given you that information, there is already a warm interest. Consider all emails received as leads for future revenue. By sending weekly, bi-weekly, or monthly emails, you're reminding an avid customer or even a one-time purchase customer of your existence. Creating beautiful, well thought out emails is key. Each email should give the recipient new information, something to get excited about, and interesting graphics. Digital marketing campaigns through email have shown to be 40 times more effective than Facebook and Twitter combined and result in 17% more revenue (Shumaker, 2019). This is powerful information.

If you're hosting a competition, sale, or event your email list should promptly hear about it. Even if you're not in a financial position to give items away, consider other experiences such as a private tour. Pairing up with other vendors is also a fantastic idea to merge interests. For example, contact a local photographer and come to an agreement. With you offering your site, and the photographer offering a free session, you're both able to reach new target audiences. Again, opportunities like this should be put into email campaigns. Allocating some of your total budget to giveaways and campaigns will ultimately grow your audience and the return will likely be much higher than the initial amount set aside.

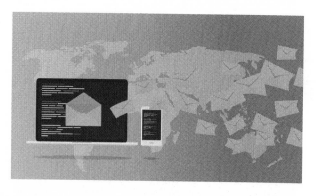

The second avenue of digital advertising is social media. Social media marketing can be divided into two categories: natural and paid. A natural following will happen as you post videos, photos, or information onto your social media. This is an area to save, as many tools like Instagram, Facebook, and

Twitter are free to use. With all of these being free, capitalize on each. Many of these outlets even have a business tool for when your following grows.

Paid social media refers to ad buys. Ad buys will be useful once you figure out your demographic. What type of person is interested in your winery? Take into consideration demographics like age, locals, visitors, tourism opportunity, males or females. Social media allows you to target a specific demographic based on those factors with more specialized ads.

The next thing to discuss is "search engine optimization" or SEO. SEO success is derived from algorithm knowledge. Having optimal SEO potential allows your website to rank higher on search engines such as Google, Yahoo, or Bing. Consider SEO from the start, because the more content you have on your website, the more time SEO optimization takes later. The factors that go into SEO are internal and external link building, keyword research, and content creation.

According to Hearst (2019), a digital media service group with over 150 years experience, "External link building is about gaining links to your website content from other reputable sites. There are several tactics that can be used to generate backlinks, some more time consuming than others, but the general rule is, the more high-quality backlinks you have pointing to your site, the higher you will rank in search, and the best way to get people to link back to your content is to create great content." Backlinks are important for your business because Google and other engines recognize the amount of backlights a site has and ranks them by popularity when keywords are searched. Reputable, popular sites that have links to your site are much more important than low traffic, smaller websites. Backlinks from reputable sites, through mentioned or articles will ultimately help your website get noticed.

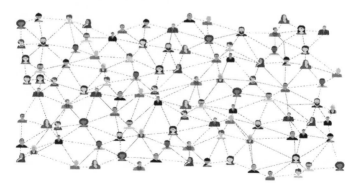

The second component to SEO is internal link building. This method takes place within your own website, but works to boost your popularity when searched. Internal link building uses topic clusters or pillar pages to create a network of information within your site. A "pillar page" will be a lengthy page of information that will link to multiple, other "topic clusters' 'or articles on your website. Each topic cluster or article must link back to your original pillar page to be effective. Your pillar page may be the home of your website and list information about your products, information on your vineyard, or information on your story. In turn, links would be provided to articles most specific, relating to those topics. Once the reader clicks on a topic page, such as information relating to your vineyard, a link back to the homepage is accessible. This is an example of effective topic clusters and pillar pages to maximize internal link building. This method helps search engines increase your ranking, which will increase traffic to your website.

Lastly, content creation will improve SEO. Content creation is related to link building in that it should always be in pillar/topic cluster style. The more content you have, the stronger the internal links will be for your website. Remember not to throw creativity and professionalism out the door. Internal linking is mostly useful when a reader sticks around and wants to actually click on the links to read further. Existing content can always be improved and edited to create more links. These require minimal effort and will get your website results.

In terms of budget and the three factors of SEO, time is the resource most needed. Be intentional with these practices as they do not require a large investment and will change the traffic on your website. If you do not wish to spend the time, outsource because social media will grow your brand. More traffic on social media leads to more customers.

Chapter 6: Seven Common Mistakes Wineries Make & Solutions

This chapter is meant to run through the most common mistakes a winery can make. Luckily, most of the solutions are already in this book. While there are many mistakes a winery can make, these are the ideals that should be learned right away. If you're a beginner, mistakes are bound to happen. Hopefully, after reading this guide those mistakes will be small and you'll be able to learn and move forward. This chapter will outline some of the bigger mistakes, and bring attention to key ideals. Mastering these concepts from the get-go will set you up for success in the long run.

1. No Business Plan and Inadequate Records

One of the most common mistakes a winery makes is not having a solid business plan. A business plan encompanies all aspects of the business including financial costs and goals. Strategy, planning, a clear vision, and budgets are needed to create a solid foundation before getting started. Often times vineyards and wineries will focus too much on the wine, and not on the concepts of running a business.

Many focus on how they will make the wine when the real question is, "how will I sell the wine once it's made?" Having and making wine does not ensure financial success. With so much competition, wineries need to have a solid plan put in place so the marketing is solid. Rising costs, economic cycles, and consolidation trends should be paid attention to as much as the winemaking process itself. Before you invest in a winery, it's important to think about pricing, research the market, sort out financials, and create goals.

The second area where wineries make a mistake is not taking the time to create accurate documentation. Often a stellar wine will be crafted (perhaps even by chance) and cannot be recreated or duplicated because of inadequate records. Because grapes are influenced by an array of components, it's important to keep track of all the elements. Weather patterns, temperatures used during the fermentation process, the amount of additives, and yeast quantities should all be documented for future years. On the other hand, if something goes wrong and the wine is spoiled, thorough documentation will tell you what went wrong and how to avoid it in the future. Good, accurate records should be carried out daily. Log any wine treatments, analyses, tasting notes, and more.

2. Harvest

One of the most common rookie mistakes made in the vineyard is harvesting at the wrong time. If you harvest too early, you may be stuck with undesirable notes, bitterness, and undeveloped flavors. If you harvest too late, your delicate blends may become flabby and/or too high in alcohol levels. Monitoring the grapes and upcoming weather patterns is imperative. Be sure to take note of Brix, pH, and TA. Many winemakers will use their taste buds to determine the state of the grapes. Around harvest this should be done regularly and even daily once harvest is near. Tasting the grapes can help evaluate the skins thickness, texture, intensity of the flavors, and quality of the tannins. As you gain experience you will be able to make inferences based on the seed color and berry texture. There should be a desirable balance between the sugar, acidity, tannins, and flavor compounds.

An early harvest may be necessary if weather events loom. Heat waves, long periods of rain, and frost can ruin a crop quickly. If this is in the forecast, consider an early harvest. Leaving the grapes for too long may result in losing the entire crop or unbalanced grapes. Remember that each year may be different and a set date for harvest rarely works year after year. Most varieties will have a different harvest period. Wines with higher acidity must be harvested earlier, while sweeter wines will have a longer growing season. In addition, grapes in different areas of the vineyard will be subject to different amounts of light and temperatures. Decide whether to harvest the whole bunch, or hand pick grapes as they ripen. In most cases, not all of the grapes are harvested at once due to variations on the plot. Be sure to study each grape and keep all of these various components in mind to produce the best quality final wine.

3. Additives vs. Quality

While there are many additives that can be added while processing wine, many rookie vintners will make the mistake of adding too many. The most popular additives include acid, water, enzymes, nutrients, tannins, bentonite, and sulfur dioxide. Rookie vintners make the mistake of thinking that the more tweaks and additions they make, the better the final product will be. However, experienced winemakers know that additives are only necessary when the grapes truly need it. Many additives when used correctly really aren't additives at all and instead glom onto unwanted particles for easier removal.

There are many examples where additives may be necessary and should not be confused with unnecessary manipulation. For example, sulfur is typically added to kill unwanted bacteria and yeast in the winemaking process. When sulfur is used it must be disclosed if the wine contains over ten parts per million in the final product. There are a small group of individuals who are sensitive to sulfur, but most individuals are not. Yeast is another microorganism that's important for the winemaking process. Yeast turns sugar into alcohol and is done naturally. Rookie winemakers will create ambient yeast and cultured yeast and add them to the wrong variety.

Understanding your additives and knowing when they are necessary is important. Another possible additive is sugar. While some winemakers will use this method to increase the final alcohol levels, rookie winemakers may do this unnecessarily, tainting the wine. Adding sugar which is referred to as chaptalization is also illegal in many areas of the world, including California.

Bottom line - growing quality grapes in a simple manner is more beneficial than using additives. Producing quality grapes starts early on in the process, making it imperative that each decision aligns with producing quality yields. Your philosophy should be about producing such great grapes, where additives are rarely needed.

4. Sanitation

Another common mistake made by new vintners is improper sanitation of the winery. This is not only referring to surface level, common cleaning techniques, but also at the cellular level. The first reason sanitation is important is quite obvious. Consumers expect safe products and various government agencies ensure this to the customer. This system starts with you. So not only is it important, but to stay in business you'll need to follow the guidelines put forth by various agencies. Wine can pose a serious risk when contaminated. Maintaining a clean and sanitary winery from the start is a good business practice. A clean working environment encourages morale, efficiency, equipment longevity, and consumer confidence. Having good sanitation practices boosts the image of your business.

Second, having good sanitation and cleaning practices will rid the possibility of tainted wine. Remember that wine gets its color from grape to skin contact. If white must and red must are in contact, white white will no longer be white and take on color. Every machine should be cleaned when handling different varieties through equipment.

During the winemaking process, there are multiple pieces to consider for sanitation. For example, stoppers airlocks, tubing, bottles, hydrometers, fermenting containers, wine thiefs, measuring equipment, brushes and more should all be sanitized. Basically anything that comes in contact with the wine should be cleaned thoroughly.

The process of sanitation begins with a clean water rinse with a detergent. The surfaces are then rinsed with water. Most vintners will then follow this process with an acid rinse followed by an additional water rinse. Hot water is not always used or needed and can be dangerous when mixed with the solvents used. So be sure to read labels and learn how the product you've chosen works. The best sanitation products are considered to be Sodium metabisulfite, Trisodium phosphate, and citric acid. Bleach and chlorine should be avoided.

5. Photos and Apps

In today's world it's not always enough to sell quality wine. Consumers want to interact and get to know the wineries from which they purchase from. While there are many aspects to having a strong online presence, this portion will discuss the important areas some wineries miss the mark.

Often consumers will look up reviews and decide how to spend their time after seeing photos of a property. If there are only two wineries in your location, one having beautiful photos and the other not, the winery with the photos will definitely get more traffic. A picture is worth a thousand words. Visitors want to know the potential of a vineyard before visiting in most cases. Be sure to take dozens of beautiful photos not only for your website but for your social media pages. Sites like Yelp and Google reviews should also be developed with numerous photos. In addition, encourage customers to create and share their own photos. This takes some of the workload off of you. Having individuals visit your property and post videos or photos is a win win.

According to Brandwatch (2019), ninety percent of Instagram users are under the age of thirty-five, and the platform now boasts 800 million users. Engagement with brands on Instagram is ten times higher than on Facebook. Instagram is another important tool to be developed to reach one of your target audiences. Because many individuals have an

Instagram account, growing your following and increasing your shares will increase visits to your vineyard. To increase your engagement on your socials, hold a photo contest every month. At the end of the month choose the best ten photos you're tagged in and have your followers decide who wins. This gives incentive to participate and will boost your popularity.

The second mistake wineries make is not exploring or taking advantage of liquor delivery apps. Although not available in some locations, apps like Drizly, Minibar Delivery, Saucey, Kink, Swill, and Amazon Prime Now are growing in popularity. Look into these modern apps and see if they're available in your area. This is a simple avenue to increase sales and market to the younger generations who are tech/app savvy.

6. Not Filling Vessels

When making wine, it's extremely important to fill every vessel to the top. This includes carboys, demijohns, and fixed capacity tanks. When tanks are not filled to the top, empty space is left in the vessel. Air space will cause oxidation which can spoil wine. When using barrels, keep in mind that a small amount of oxidation will reoccur and is actually helpful for the wine. This is called micro-oxidation. Barrels that promoto micro-oxidation will create a creamier feel, and better expression of fruit notes. However, barrels typically need to be topped off with additional wine to prevent total oxidation and too much evaporation.

7. Ignoring or not Creating Brand Identity

A brand is different than a name or label. A brand collects many different aspects and produces an overall image. For a winery to receive extra recognition, the customer must be able to identify with the brand as a whole. You shouldn't have your website in one style and the label of your bottles in another. For example, don't use a cityscape on your bottle and come off one way, then use romantic images on your website and come off another. Every photo, font, and overall visual style should be cohesive. Logo, label and website design should be cohesive and tell the same story with minor variations differentiating them. Figure out your story and stick to it.

The first step to creating brand identity is to first define your business. What is important to you, what are your values, what do you hope to portray to customers? Create a mission statement and vision. This should be solid before the brand launches. The second step to creating brand identity is to identify your target audience. Ask yourself what this audience expects and wants from your product. For example, what decisions do you need to make to make your brand more appealing to that audience? Catering to consumers doesn't mean you need to change everything you stand for. It means you have to expose your values in a way that resonates with them. The third step is to determine your unique qualities and how you can market them for others to notice. If you have a specialty or unique topography situation that plays a role in your wine, how can you portray that to customers in an appealing way? Figure out your strong suits and how you will highlight them when it comes time to sell. Lastly, make a plan for your visual materials. This included website design, logos, labels, bottle design, press materials, and more. Every aspect needs to look like it came from the same winery, but still be unique and spark interest in its own way.

Establishing your brand will make your business more memorable. Customers who establish brand loyalty will be of benefit to you for future sales. The main takeaway is to not confuse the customer, and create a clear vision recognizable to all.

Chapter 7: Accounting and Smart Tax Practices

This chapter will briefly touch on accounting and tax practices. Because each winery is subject to state and federal laws, professional help should be enlisted to be sure your business complies. When setting up a vineyard, business wise, one of your investments should be in a good business lawyer and accountant. Unless you are licensed in both, there will be a lot of gray areas that should be looked over by a professional. This chapter is to give you a rough idea on some of the accounting and tax methods involved with your business.

Accounting

Doing an analysis of your business will help you understand the potential growth your vineyard has. Determine costs, calculate benefits, compare alternatives, and create a report/plan of action to have a better business. The major areas to consider for costs can be divided into four categories: vineyard, winemaking, cellar, and packaging. Most costs will likely fall under one of those categories. To develop a cost model, make a list of potential costs. The first aspect to account for are materials. Begin by tracking the prices for different grape varieties. Harvesting or purchasing grapes will lead to different costs and will look different on a cost model. This is because the cost of harvest is different than the cost of purchase, advantages and disadvantages aside.

After harvest, when the grapes are brought to the winery, the new quantity should be documented. Meaning, although you have x amount of grapes from harvest, after being pressed or crushed you now have x amount left. Next, your model should take into account packaging materials such as bottles, corks, cases, and labels. After documenting all material costs, start considering are landed costs, labor, and overhead. Landed costs include freight, duties, and fees. Landed costs must be evaluated to get the true costs of good.

For example if outsourcing grapes, additional costs may arise with transportation or perhaps taxes/duties. The main purpose of landed costs are to find the true value of a good, taking all aspects surrounding that good into consideration. Labor costs should include the time and expenses put in from the beginning of the process to the end. Each stage of the winemaking process should be documented in terms of labor. The last area to be looked at when building a cost model is overhead. This includes but is not limited to chemicals used, fertilizers, building rent, insurance, utilities, maintenance, and storage. Once these costs are documented, an experienced accountant or software program can give helpful predictions and information.

Aging costs are associated with costs after the wine is pressed. When wine is pressed, as you know, there are many more steps until it can be bottled. Labor, maintenance, storage, utilities, and evaporation are aging costs to be evaluated.

The major areas of cost drivers and savings is freight. This includes every time inventory is touched or moved. Each touch or time spent will likely produce an extra cost. Freight included costs for acquiring grapes, transporting grapes, how much time is spent, and physical labor. Saving potential comes in when any of these areas can be improved. For example, to save on freight, research if you are using the best priced carrier. Decide if outsourcing is necessary or if moving some freight operation inhouse may be beneficial. Evaluate if any of your practices can be changed in terms of freight to avoid any fees the winery is taking on.

Another area to evaluate is product loss which can occur each time the grapes are moving. This mainly has to do with outsourcing. Emptying, filling up new tanks, or moving the product should only be done when necessary to avoid losing product. If you are outsourcing at any step, evaluate how much that is costing versus doing it yourself to save.

Labor costs are the third area to evaluate in terms of cost drivers and saving potential. Paying employees is important. However, if your labor costs are soaring year after year, and revenue is stagnant, it's time to consider other options. Adjusting your work force may be necessary over the years to maximize revenue. Understanding labor costs and knowing if it's adding to the value of the final product is important.

The last area to evaluate when discussing saving potential is maintenance. Add up the costs of machinery and document them over the months and years. Figure out why those costs are necessary and what can be done to minimize them. For example, if equipment failure is capturing a large part of revenue, determine the cause. Old equipment may need to be replaced and the investment will save money over time.

The reason for gathering an abundance of information is to determine profitability. Once the information is gathered and accurate, it's time for review. Each wine variety being produced should be evaluated separately, but compared to your top seller. Calculate the differences and ultimately decide how much the lower revenue producing wine is costing to produce.

Tax Opportunities

Wineries are at an advantage and with a few key practices you will be able to reduce your tax exposure. The following list will showcase some of the areas your vineyard may be able to save on taxes. Because taxes are unique to your location, a tax advisor should be contacted for relevancy and more potential saving areas.

1. Entity Advantages

The entity type chosen will ultimately determine your position on taxes. C corporation structures requiring double taxation are often aren't commonly chosen. LLC's are typically the most popular amoung vineyards but S corporations are rising due to their benefits. With the Affordable Care Act, the flow-through income from an S corporation to an active shareholder is not classified as self-employment income. As a result, S Corporations aren't subject to self-employment tax or additional Medicare tax from laws put into effect in 2013. However, LLCs provide flexibility in allocating losses to family members that helped fund the winery. This allows them to offset their taxable income by using the allocated losses associated with the winery. This method can be helpful as the first five years of the vineyard may only generate losses. For example, if a winery is owned by two partners and partner A contributed 100 percent of capital, the losses can be allocated to partner A (if the business is an LLC). On the other hand, if the winery is an S corporation, losses are allocated according to the amount of shares owned by each shareholder. This scenario doesn't take into account who funded the business. For example, if two partners are equal owners, shareholder B will be allocated fifty percent of losses that are able to be deductible only to the extent shareholder B has basis.

The end goals you decide on should also play a role in establishing your business. How long will your business be around for? Do you plan on selling it in the future? What's your exit strategy? Will you pass it on to anyone? An LLC may have benefits in concern to estate planning but another entity may give benefits for outside sales or transfering existing licenses. An attorney and CPA will be able to go over those disadvantages and advantages.

2. Deductible Expenses

If you have an intention to make a profit, vineyard owners can deduct ordinary and necessary expenses under farm/business expenses. Many farming specific, operational, and administrative costs qualify as ordinary and necessary farming expenses. Agricultural businesses also have the ability to deduct more expenses than most other businesses.

3. Depreciation

Because the IRS has many methods for depreciating property, tax breaks may be available. IRS Code Section 179 and Section 168K give information of the depreciation advantages that may be available. It's advised to check with your state regulations as they may not comply with those IRS sections. Vineyards typically make large investments in different types of assets like real estate, farm equipment, machinery, and plot development. In regards to grapes, those assets may include trellis systems, irrigation systems, rootstock, vines, fences, roads, wells, and drainage for example.

4. Net Operating Losses (NOLs)

A NOL can be filed if deductible expenses are greater than income for the annual quarter. NOLs can be evaluated for the previous two years, or take into consideration the next twenty. It may be possible to carry back farming loss to receive a refund of taxes paid in past years. If a NOL is not carried back, it can be carried ahead to offset future revenue. Your state laws should be evaluated to see if this provision can apply to you, as not all states conform to NOL rules put forth by the IRS.

5. Cash Method and Accrual

Deciding your accounting method, cash or accrual, is important. Vineyards (as farming operations) may be allowed to use the cash method regardless of size. The activity must be held by a sole proprietor or by a flow-through entity. However, a vineyard may not be able to use this method if thirty-five percent of the losses from the activity are allocated to limited partners or limited entrepreneurs. If a vineyard is a C cooperation, thresholds may also preclude the vineyard from the cash method. In some cases there may be a small-taxpayer exception. To meet this requirement, the gross income for the past three years must be less than one million. If your business can meet this, the cash method may be worth looking into. In this circumstance, grapes, glass, corks, and other materials may be considered raw materials which gives more breaks. If you're able to use the cash method, your vineyard may be able to benefit more from capitalizing or "expensing" preproductive costs. The cash method allows such costs to be deducted until the year they're paid which reduces taxable income promptly. It advised that you speak with your lawyer, CPA and accountant for more specific, regional tax information and possible opportunities.

6. Assets in Real Estate

It's possible for vineyards to have considerable real estate holding to borrow against. This is due to the capital investment that is sometimes necessary for vineyard activities.

7. Soil and Water Conservation

Expenses in regards to soil and water conservation may be deducted and not capitalized. Information surrounding this topic is required by the government to protect endangered species and give valuable information to agencies.

8. Post Harvest/Pre-Bud Break Costs

If using the accrual method, post-harvest and pre-bud break costs may be able to be deducted.

9. Vineyard Appellation

Land costs in most cases can not be depreciated or amortized. However, there is a provision that applies to the growing of grapes. It may be possible to segregate an AVA Designation as an intangible asset. It's possible it could then fall into amortization.

10. Taxpayers as Farmers

If you are a taxpayer that qualifies as a farmer, there could be more favorable rules that apply to the payment of estimated taxes.

11. Income

It's possible you may qualify to use farm income averaging. This is the ability to take the average of current year's farm income or only some of the costs, and distribute it out over the past three years. Having income spread over the past three tax years can mitigate the tax impact for higher-earning years. This can help vineyards stay out of the top tax bracket for a high-earning year. If you operate a vineyard and a winery, be sure to determine the exact amount of income relating to farming. Tax rates are higher now than in the past, and this is one provision many can take advantage of.

12. Fuel Tax Credit

Fuel used for farming purposes may be subject to a tax credit or refund of federal excise taxes.

13. Williamson Act

The Williamson Act is an act that reduces property taxes for some qualifying properties. It should be looked into to see if your personal region could fall into this category.

14. Savings on Agricultural Items

Some farm equipment, machinery and other items may be exempt from sales tax.

Conclusion

Starting a winery business can be a great decision in terms of profitability. Although there are many wineries and choices for consumers, the consumer is always searching and willing to try new wine. This is unique to the wine business because the curiosity around wine is ever growing with no plans to stop. Although wineries take a lot of time, effort, and planning the possible revenue available to you is worth it.

After reading this guide, you now know the basics of the wine industry. The industry and terminology associated is a whole new language in of itself. Being able to converse with customers, potential buyers, and other vintners will improve your business. It's important to understand and appreciate the complexities that go along with making wine so when something goes wrong, you'll have the knowledge to fix it. After reading this guide, you're well equipped with the information necessary to build a sold business plan. Potential cost examples were given, but now it's time to put those principles into action. If you haven't already, use this guide to make the first draft of your business plan. Potential costs, goals, legalities, and insurance should all be put into place.

In addition, you know now all of the components that go into a wine's final taste. From site selection, to location, to soil, vineyard design, and overall practices, these decisions are now up to you. After reading about the various wine regions in the United States, you should have a better idea of what grapes will grow best in your area. Vineyards have successfully grown wines for years now and there is no reason why your business can't flourish in the same way! States in the US all have a different approach to growing grapes based on their location. Studying their successes will put you on the right path and give more time to ponder how you will make your product stand out. Choose a grape variety that already works well with your conditions in terms of elevation, soil, weather patterns, and growing season. This will minimize costs as less problems are likely to arise.

Once you're ready to go in terms of selection, it's time to grow the grapes. This guide has given you everything you need to know to get started. From preparing the soil, to bottling the wine, we've got you covered. In addition, it's important to know the basic processes and know where to put your spin on things. Creating a wine unique to your winery will help you stand out and gain mention. Many examples have been given with helpful ranges included. The final decisions are up to you and should be determined based on your preferences that will develop as you gain experience.

Along with knowing how to produce wine, the selling aspect should always be the main focus. After reading this guide, there should be no doubt that you can produce something excellent. However, even excellent wines can slip under the radar if the marketing is not adequate. Remember all of the options and ideas given that will help you sell more bottles. From opening a tasting room, to holding events, wine clubs, and influencers, these techniques are sure to give your business the attention it deserves. Marketing online is also discussed in this guide and is a valuable resource that needs to be taken advantage of. Remember to create a website that will catch the eye of consumers and stay strong on your social media pages. Refer back to the guide to learn how to maximize traffic and interest.

Lastly, remember that all wineries make mistakes but they don't have to be detrimental. After reading this guide you should have a great understanding on how to minimize problems and avoid the mistakes many wineries make. Methods on accounting and tax opportunities should be discussed with an accountant or lawyer for guidance. We really hope you enjoyed this guide and found it useful. Remember to refer back to the guide as you go, and we're confident your business is bound for success!

References

8 Steps to Owning Your Own Vineyard. (2008, January 1). Retrieved from https://www.inc.com/ss/8-steps-to-owning-your-own-vineyard

50 Incredible Instagram Statistics You Need to Know. (2019, January 20). Retrieved from https://www.brandwatch.com/blog/instagram-stats/

Conway, J. (2019, August 9). Number of wineries in the U.S. by state, 2019. Retrieved from https://www.statista.com/statistics/259365/number-of-wineries-in-the-us-by-state/

Fickle, L. A., Folwell, R., Ball, T., & Clay, C. (2015, February). Agribusiness Management. Retrieved from http://www.agribusiness-mgmt.wsu.edu/AgbusResearch/SmInvestWinery.htm

Franson, P. (2012, September 13). Retrieved from https://winesvinesanalytics.com/news/article/105157/Growing-Grapes-by-the-Numbers

Franson, P. (2016, May 12). Tasting Room Survey Discloses Disparities. Retrieved from https://winesvinesanalytics.com/news/article/1687 37/Tasting-Room-Survey-Discloses-Disparities

Frazier, R. (2017, February 23). Marketing Budget: How to Set One for Your Business (Updated). Retrieved from https://www.visigility.com/marketing-budget/

Goldstein, C. (2020, January 31). How to Start a Winery: 5 Steps to Starting a Wine Business. Retrieved from https://www.fundera.com/blog/how-to-start-a-wine-business#step4

Guide to California Wines. (2019, September 25). Retrieved from https://www.marketviewliquor.com/blog/2019/09/ guide-to-california-wines/

Hagen, W. (n.d.). Planning Your Backyard Vineyard. Retrieved from https://winemakermag.com/article/561-planning-your-backyard-vineyard

Haibach, R. (2018, August 27). Target Sugar and Acid Levels for Popular Wine Grape Varieties. Retrieved from https://www.smartwinemaking.com/post/optimal-sugar-and-acid-levels-for-popular-wine-grape-varieties

How to Start a Vineyard Business. (2019, December 9). Retrieved from https://howtostartanllc.com/business-ideas/vineyard#useful-links

Lee-Sedgwick, H. (2018, April 8). What is a Sommelier? Retrieved from https://www.winecountry.com/blog/what-is-a-sommelier/

Margalit, Y. (2003). Winery technology & operations: a handbook for small wineries. San Francisco, CA: Wine Appreciation Guild.

Moon, A. (n.d.). What Seasons Do Grapes Grow? Retrieved from https://www.hunker.com/12548730/what-seasons-do-grapes-grow

Pinsker, J. (2014, October 1). The Psychology Behind Costco's Free Samples. Retrieved from https://www.theatlantic.com/business/archive/2014/10/the-psychology-behind-costcos-free-samples/380969/

Puckette, M. (2019, September 10). Understanding Acidity in Wine. Retrieved from https://winefolly.com/deep-dive/understanding-acidity-in-wine/

Robinson, J. (2003). Jancis Robinsons wine course: a guide to the world of wine. New York: Abbeville Press Publishers.

Shumaker, R. (2019, May 27). Digital Marketing for Wineries: Where to Invest Your Budget. Retrieved from https://marketing.sfgate.com/blog/digital-marketing-for-wineries-where-to-invest-your-budget

Stafne, E. (2019, June 20). Vineyard Design. Retrieved from https://grapes.extension.org/vineyard-design/

Tebib, K., Besançon, P., & Rouanet, J. M. (1994, December). Dietary grape seed tannins affect lipoproteins, lipoprotein lipases and tissue lipids in rats fed hypercholesterolemic diets. Retrieved from https://www.ncbi.nlm.nih.gov/pubmed/16856327

Thomson, A. K. (2019, July 24). Why tech workers are not buying traditional California vineyards. Retrieved from https://www.ft.com/content/621f7abe-a947-11e9-90e9-fc4b9d9528b4

Top 15 Wine-Producing Countries. (2019, November 8). Retrieved from https://italianwinecentral.com/top-fifteen-wine-producing-countries/

Welch's. (2020). Retrieved from
https://www.grapediscoverycenter.com/welchs

Wine Regions in California. (2018, October 5).
Retrieved from
https://usawineratings.com/en/blog/insights-1/wine-regions-in-california-46.htm

Made in the USA
Middletown, DE
19 July 2020